Jens Boy

External drivers of biogeochemical cycles in tropical Andean forest

Jens Boy

External drivers of biogeochemical cycles in tropical Andean forest:

the impact of extreme climatic events and biomass burning on a megadiverse ecosystem

Südwestdeutscher Verlag für Hochschulschriften

Impressum/Imprint (nur für Deutschland/ only for Germany)
Bibliografische Information der Deutschen Nationalbibliothek: Die Deutsche Nationalbibliothek verzeichnet diese Publikation in der Deutschen Nationalbibliografie; detaillierte bibliografische Daten sind im Internet über http://dnb.d-nb.de abrufbar.
Alle in diesem Buch genannten Marken und Produktnamen unterliegen warenzeichen-, marken- oder patentrechtlichem Schutz bzw. sind Warenzeichen oder eingetragene Warenzeichen der jeweiligen Inhaber. Die Wiedergabe von Marken, Produktnamen, Gebrauchsnamen, Handelsnamen, Warenbezeichnungen u.s.w. in diesem Werk berechtigt auch ohne besondere Kennzeichnung nicht zu der Annahme, dass solche Namen im Sinne der Warenzeichen- und Markenschutzgesetzgebung als frei zu betrachten wären und daher von jedermann benutzt werden dürften.

Verlag: Südwestdeutscher Verlag für Hochschulschriften Aktiengesellschaft & Co. KG
Dudweiler Landstr. 99, 66123 Saarbrücken, Deutschland
Telefon +49 681 37 20 271-1, Telefax +49 681 37 20 271-0, Email: info@svh-verlag.de
Zugl.: Mainz, Gutenberg Universität, Diss., 2009

Herstellung in Deutschland:
Schaltungsdienst Lange o.H.G., Berlin
Books on Demand GmbH, Norderstedt
Reha GmbH, Saarbrücken
Amazon Distribution GmbH, Leipzig
ISBN: 978-3-8381-0674-8

Imprint (only for USA, GB)
Bibliographic information published by the Deutsche Nationalbibliothek: The Deutsche Nationalbibliothek lists this publication in the Deutsche Nationalbibliografie; detailed bibliographic data are available in the Internet at http://dnb.d-nb.de.
Any brand names and product names mentioned in this book are subject to trademark, brand or patent protection and are trademarks or registered trademarks of their respective holders. The use of brand names, product names, common names, trade names, product descriptions etc. even without a particular marking in this works is in no way to be construed to mean that such names may be regarded as unrestricted in respect of trademark and brand protection legislation and could thus be used by anyone.

Publisher:
Südwestdeutscher Verlag für Hochschulschriften Aktiengesellschaft & Co. KG
Dudweiler Landstr. 99, 66123 Saarbrücken, Germany
Phone +49 681 37 20 271-1, Fax +49 681 37 20 271-0, Email: info@svh-verlag.de

Copyright © 2009 by the author and Südwestdeutscher Verlag für Hochschulschriften Aktiengesellschaft & Co. KG and licensors
All rights reserved. Saarbrücken 2009

Printed in the U.S.A.
Printed in the U.K. by (see last page)
ISBN: 978-3-8381-0674-8

Contents

Contents I
List of tables IV
List of figures VI
List of abbreviations IX
Summary X
Resumen XI
Zusammenfassung XIII
Acknowledgments XV

A   Summarizing overview 1

1. Introduction 2

2. Materials and Methods 5
    2.1. Study site 5
    2.2. Field sampling 8
    2.3. Hydrological measurements 9
    2.4. Chemical analyses 10
    2.5. Remote sensing and transport pathway reconstruction 11
    2.6. Calculations and statistical analyses 13

3. Results and Discussion 17
    3.1. Tropical Andean forest derives calcium and magnesium from
        Saharan dust (Section B, pp 39-60) 17
    3.2. Amazonian biomass burning derived acid and nutrient deposition in the
        north Andean montane forest of Ecuador (Section C, pp 61-86) 19
    3.3. Water flow paths in soil control element exports in an Andean tropical
        montane forest (Section D, pp 87-124). 20
    3.4. Integrative statistical evaluation of the controls of Ca and N fluxes from
        rainfall through the forest ecosystem to surface runoff 22
    3.5. Error discussions 31

4. Conclusions 32

5. References 34

B Tropical Andean forest derives calcium and magnesium from Saharan dust **39**

  1. Abstract *40*

  2. Introduction *41*

  3. Materials and Methods *42*
    3.1. Study site 42
    3.2. Field sampling 44
    3.3. Hydrological measurements 45
    3.4. Chemical analyses 46
    3.5. Calculation of deposition rates and element fluxes 46
    3.6. Remote sensing and transport pathway reconstruction 47

  4. Results and Discussion *49*
    4.1. Base metal input and response of the ecosystem 49
    4.2. Source identification 51
    4.3. Relation of precipitation and Saharan dust transport across Amazonia 55

  5. Conclusions *56*

  6. Acknowledgments *57*

  7. References *57*

C Amazonian biomass burning derived acid and nutrient deposition in the north Andean montane forest of Ecuador **61**

  1. Abstract *62*

  2. Introduction *63*

  3. Materials and Methods *64*
    3.1. Study site 64
    3.2. Field sampling 65
    3.3. Hydrological measurements 66
    3.4. Chemical analyses 67
    3.5. Calculation of deposition rates and element fluxes 68
    3.6. Remote sensing and transport pathway reconstruction, and definition of biomass-burning season 69

  4. Results and Discussion *71*
    4.1. Influence of Amazonian biomass burning on element deposition in the north Andes 71
    4.2. Effect of deposited elements on the nutrient budgets of the Andean montane forest 75

  5. Conclusions *83*

  6. Acknowledgments *83*

7. References..83

D Water flow paths in soil control element exports in an Andean tropical montane forest..87

1. Abstract...88

2. Introduction..89

3. Materials and Methods...91
    3.1. Study site..91
    3.2. Field sampling..92
    3.3. Hydrological measurements...94
    3.4. Water analyses..95
    3.5. Soil analyses...96
    3.6. Calculations and statistical evaluation...96
    3.7. Modelling of dissolved metal speciation..99

4. Results and Discussion...99
    4.1. Hypothesis 1: The concentrations of chemical constituents in stream water of steep, forested catchments are related to discharge levels in a way that is specific for each chemical constituents........................99
    4.2. Hypothesis 2: The depth of water flow in soil determines the concentration of chemical constituents in stream water...................106
    4.3. Hypothesis 3: Discharge level classification is a suitable tool to estimate the contribution of different flow regimes to element export from steep, forested catchments for long time series.........................115
    4.4. Hypothesis 4: Storm events have a significant influence on catchment nutrient export...118

5. Conclusions..119

6. Acknowledgments..121

7. References..121

E Appendix..125

List of tables

Table A-1: Means and ranges (in brackets) of selected soil properties in an Ecuadorian lower montane forest (n=47).. 7

Table A-2: Results of general linear models with a) Ca concentrations, b) Ca fluxes, c) N_{tot} concentrations, and d) N_{tot} fluxes in rainfall as target variables... 32

Table A-3: Results of general linear models with a) Ca concentrations, b) Ca fluxes, c) N_{tot} concentrations, and d) N_{tot} fluxes in throughfall as target variables... 25

Table A-4: Results of general linear models with a) Ca and b) N_{tot} concentrations in litter leachate as target variable..........27

Table A-5: Results of general linear models with a) Ca and b) N_{tot} concentrations at 0.15 m soil depth and c) Ca and d) N_{tot} concentrations at 0.3 m soil depths as target variables...29

Table A-6: Results of general linear models with a) Ca and b) N_{tot} concentrations in runoff as target variables.............30

Table B-1: Mean annual fluxes of Ca, Mg, and K [kg ha^{-1} a^{-1}] and standard deviations of rainfall, throughfall, dry deposition and canopy budget of the five hydrological years of the three microcatchments (MC1- MC3)...............50

Table C-1: Comparison of annual bulk deposition (kg ha-1 yr-1) with rainfall in selected tropical montane forests..72

Table C-2: Water fluxes of rainfall (RF), throughfall (TF), stemflow (ST), litter leachate (LL), soil solution at the 0.15 m and 0.3 m depths in mineral soil (SS15 + 30), and stream water (SW) of an Andean montane forest during biomass burning in Amazonia ("fire") and without ("no fire") between May 1998 and April 2003 for the five transects (T1- T3)................................75

Table C-3: Dry deposition at an Andean montane forest during biomass burning in Amazonia ("fire") and without ("no fire") between May 1998 and April 2003 for the five transects (T1- T3). Different lower case letters indicate significant differences of mean concentrations between "fire" and "no fire" (Games Howell p<0.05)..76

Table C-4: Volume-weighted mean concentrations in litter leachate (LL), soil solution at the 0.15 m and 0.3 m depths in mineral soil (SS15 + 30), and stream water (SW) of an Andean montane forest during biomass burning in Amazonia ("fire") and without ("no fire") between May 1998 and April 2003 for the five transects (T1- T3). Different lower case letters indicate significant differences of mean concentrations between "fire" and "no fire" (Games Howell p<0.05)...81

Table D-1: Occurrence and seasonal distribution of measured and modeled discharge levels over the five flow classes of three small streams draining microcatchments (MC 1-3) under montane rain forest in southern Ecuador (April 1998-April 2003).........100

Table D-2: Mean 5-year discharge for the different flow classes and MCs, used as q in the flow class effect (FE) calculation..103

Table D-3: Mean concentration of C, N, P, and S in stream water and flow class effect (FE) of the different flow classes for the three streams. Different lower case letters indicate significant differences among flow classes (Games-Howell, $p<0.05$)..105

Table D-4: Mean concentration of K, Ca, Mg, Na, Al, and Mn in stream water and flow class effect (FE) of the different flow classes for the three streams. Different lower case letters indicate significant differences among flow classes (Games-Howell, $p<0.05$)..108

Table D-5: Flow-weighted mean (fwm) concentration, "conventional" (fwm-based) export, modeled export, and difference between conventional and modeled export of C, N, P, and S of the three microcatchments between April 1998 and April 2003..110

Table D-6: Flow-weighted mean (fwm) concentration, "conventional" (fwm-based) export, modeled export, and difference between conventional and modeled export of K, Ca, Mg, Na, H, and Mn of the three microcatchments between April 1998 and April 2003..112

Table D-7: Mean element export of C, N, P, and S by discharge in different flow classes (April 1998- April 2003). Percentage values indicate the fraction of total export for the corresponding flow class...116

Table D-8: Mean element export of K, Ca, Mg, Na, Al, and Mn by discharge in different flow classes (April 1998- April 2003). Percentage values indicate the fraction of total export for the corresponding flow class....................118

List of figures

Figure A-1: Location of the study area..6

Figure A2: Illustration of the factor loading of element concentrations in rainfall between May 1998 and April 2003 in the coordinate system of principal components (PC) 1 and 2. Circles indicate element grouping...............15

Figure B-1: Sahara dust transport corridor across Amazonia. Location of the 24 meteorological stations of INPE (solid circles), the three defined passage zones (shaded, Z1-Z3) and the study site (open star). As an example, daily HYSPLIT-backward trajectories (3000, 4000, and 6000m target heights, coming in from the western Sahara) are drawn as grey lines for the week of 01-07 May 1999. The transported aerosol experienced good transport probability across Amazonia (Figure B-3) and was washed out at our study site (126 mm precipitation in that week [01-07/May/1999]). There was no biomass burning in the transport corridor (as detected by the satellite NOAA-12). The Ca total deposition at the Ecuadorian montane forest for that week (01-07 May 1999) was 3.6 kg ha^{-1}..................................48

Figure B-2: Mean annual input by deposition, output by surface flow and microcatchment budget of (a) Ca, (b) Mg and (c) K of three microcatchments of tropical montane rain forest in south Ecuador between May 1998 and April 2003. Negative values of microcatchment budget indicate loss, positive values accumulation of base metals in the ecosystem. Error bars indicate standard deviations among the three catchments..52

Figure B-3: Upper graph: Temporal course of precipitation at the study site (columns), the number of daily HYSPLIT backward trajectories per month which originated in the Sahara and reached the study site at noon at the target heights 3000, 4000 or 6000 m a.s.l. (blue line), and the monthly mean TOMS aerosol index (AI) of the west African source region (35.5°N-9.5°N, 17°W-11°E, brown line). Central graph: Temporal course of Ca concentrations in rainfall samples (brown diamonds, inlet figure illustrates the close relation between Ca and Mg deposition), and probability of Saharan dust transport across Amazonia expressed in absolutely dry conditions (i.e., number of days per month with 0 mm precipitation along the passage of daily trajectories across the Amazonian dust transport corridor [Figure B-1], columns). Two volcanic eruptions during the monitoring period are indicated by arrows and name of volcano. Both volcanoes are located around 500 km to the N or NNE, respectively. Lower graph: Monthly aerosol emissions over the Sahara as detected by TOMS (panels A-I, corresponding letters in the central graph indicate the time covering). Between July and November 1998 precipitation data coverage across Amazonia was insufficient to determine transport probability (no data for 'Z3' of Figure B-1)......................................54

Figure C-1: Classification of biomass-burning periods using NO_2-index satellite data of GOME (Global Ozone Monitoring Experiment) in the air column above the study site (4°S 79°W ± 2°) and fire pixel count of NOAA 12 (National Oceanic and Atmospheric Administration) in Amazonia and northern South America along the daily Hysplit trajectories (radius 25 km around the actual parcel location)..*70*

Figure C-2: Temporal courses of rainfall concentrations of selected chemical species (dark line, running 3-months average) in comparison to the NO_2-index of GOME (light line) as indication of fire acitivity in the Amazon basin between May 1998 and April 2003. The temporal course of Ca concentrations also represents those of Mg, which are closely correlated with the Ca concentrations (r=0.85). The box indicates a period of Saharan dust deposition in 1999/2000 which caused elevated base metal deposition.. ....*74*

Figure C-3: Total deposition during biomass burning periods ("fire") compared to normal conditions ("no fire") between May 1998 and April 2003 for the microcatchments (MC) 1-3. Error bars represent standard errors and stars indicate significance (<0.05, Games Howell)...................................*77*

Figure C-4: Net canopy budget expressed in percent of total deposition between May 1998 and April 2003 for the microcatchments (MC) 1-3. Negative values indicate net loss, positive net retention of the canopy. Error bars represent standard errors. Stars indicate significant differences between "fire" and "no fire" canopy budgets (<0.05, Games Howell; statistical test based on fluxes, not percentage of total deposition).......................................*79*

Figure C-5: Percentage of total deposition being lost/retained down to a given stratum [LL= flux in litter leachate; SS15 and SS30= fluxes at the 0.15 and 0.3 m depths in mineral soil, respectively; SW= flux in surface flow (stream water)] between May 1998 and April 2003 for the microcatchments (MC) 1-3. For each transect base saturation (BS) and cation exchange capacity (ECEC) is given in column a). Direction of loss or retention is indicated in pannel b1)...*80*

Figure C-6: Correlations between Mn total deposition and N_{tot}-canopy budget (a), and between Mn- and N_{tot} canopy budget (b) for May 1999-April 2003 (1998 no N_{tot} measured) in transect MC 2.1..*82*

Figure D-1: Schematic sketch of the five flow classes (MD5Y= mean discharge over five years)..*98*

Figure D-2: Illustration of the suitability of the throughfall criterion to represent the soil water content criterion to separate lateral flow from stormflow conditions by the example of an event between 6/2/2001 and 6/7/2001. When discharge threshold and either throughfall threshold or soil water saturation threshold are crossed, lateral flow conditions apply...........................*98*

VII

Figure D-3: Mean concentrations of TOC, N, P, and S in (i.) flow class, (ii.) soil solid phase, and (iii.) ecosystem flux between April 1998 and April 2003 in the three microcatchments (MC 1-3). Abbreviations: lateral flow (LF), storm flow (ST), intermediate (IM), base flow (BF), superdry (SD), rainfall (RF), throughfall (TF), stemflow (SF), litter leachate (LL), soil solutions at 0.15m (SS-15) and 0.30m (SS-30) and streamwater (SW). Error bars indicate standard error..*102*

Figure D-4: Estimated influence of a flow class on the concentrations of chemical constituents in stream water ("flow class effect", FE). Abbreviations as in Figure D-3. Note: FE of TOC reduced by a factor of 100..................*104*

Figure D-5: Mean concentrations of K, Ca, Mg, Na, Al, and Mn in (i.) flow class, (ii.) soil solid phase, and (iii.) ecosystem flux between April 1998 and April 2003 in the three microcatchments (MC 1-3). Abbreviations as in Figure D-3. Error bars indicate standard errors..*107*

Figure D-6: Estimated influence of a flow class on the concentrations of chemical constituents in stream water ("flow class effect", FE) for a) base metals and b) Al and Mn. Abbreviations as in Figure D-3. Note: FE of Na reduced by a factor of 2...*111*

Figure D-7: Mean contributions of organo-metal complexes to total dissolved metal concentrations in the five flow classes for the three microcatchments (MC1-MC3, left panels a-c, as computed by MinteQ in weekly resolution) and mean depth profile of pH and flow-weighted mean concentration of TOC in ecosystem fluxes (right) between April 1998 and April 2003. Given pH and TOC values in the complexation graph correspond to the indicated flow classes. Abbreviations as in Figure D-3. Error bars indicate standard errors for the TOC concentrations and minima and maxima for the pH. Standard errors of the contributions of organo-metal complexes to total dissolved metal concentrations are smaller than the symbols...*114*

List of abbreviations

AAS	=	atomic absorption spectroscopy
AI	=	aerosol index
ARL	=	'Air Resource Laborartory'
BS	=	base saturation
CFA	=	continous flow analyzer
DD	=	dry deposition
DOC	=	dissolved organic carbon
DOM	=	dissolved organic matter
DON	=	dissolved organic nitrogen
ECEC	=	cation exchange capacity
ENSO	=	El Niño Southern Oscillation
FDR	=	frequency domain reflectrometry
FE	=	flow class effect
FWM	=	flow weighted mean
GLM	=	general linear model
GOES	=	'Geostationary Environmental Satellite'
GOME	=	'Global Ozone Experiment'
HYSPLIT	=	'Hybrid Single Particle Lagrangian Integrated Trajectories'
INPE	=	'Instituto Nacional de Pesquisas Espaciais'
LEA	=	canopy budget
LL	=	litter leachate
MC	=	microcatchment
MODIS	=	'Moderate Resolution Imaging Spectrometer'
NOAA	=	'National Oceanic and Atmospheric Administration'
PCA	=	principal component analysis
RD	=	rainfall deposition
SFD	=	stemflow deposition
SST ENSO	=	sea surface temperature at ENSO region
SWB	=	soil water balance
TD	=	total deposition
TDN	=	total dissolved nitrogen
TDP	=	total dissolved phosphorus
TDS	=	total dissolved sulphur
TFD	=	throughfall deposition
TOC	=	total organic carbon
TOMS	=	'Total Ozone Mapping Spectrometer'
VWM	=	volume weighted mean

Summary

Successful conservation of tropical montane forest, one of the most threatened ecosystems on earth, requires detailed knowledge of its biogeochemistry. Of particular interest is the response of the biogeochemical element cycles to external influences such as element deposition or climate change. Therefore the overall objective of my study was to contribute to improved understanding of role and functioning of the Andean tropical montane forest. In detail, my objectives were to determine (1) the role of long-range transported aerosols and their transport mechanisms, and (2) the role of short-term extreme climatic events for the element budget of Andean tropical forest.

In a whole-catchment approach including three 8-13 ha microcatchments under tropical montane forest on the east-exposed slope of the eastern cordillera in the south Ecuadorian Andes at 1850-2200 m above sea level I monitored at least in weekly resolution the concentrations and fluxes of Ca, Mg, Na, K, NO_3-N, NH_4-N, DON, P, S, TOC, Mn, and Al in bulk deposition, throughfall, litter leachate, soil solution at the 0.15 and 0.3 m depths, and runoff between May 1998 and April 2003. I also used meteorological data from my study area collected by cooperating researchers and the Brazilian meteorological service (INPE), as well as remote sensing products of the North American and European space agencies NASA and ESA.

My results show that (1) there was a strong interannual variation in deposition of Ca [4.4-29 kg ha^{-1} a^{-1}], Mg [1.6-12], and K [9.8-30]) between 1998 and 2003. High deposition changed the Ca and Mg budgets of the catchments from loss to retention, suggesting that the additionally available Ca and Mg was used by the ecosystem. Increased base metal deposition was related to dust outbursts of the Sahara and an Amazonian precipitation pattern with trans-regional dry spells allowing for dust transport to the Andes. The increased base metal deposition coincided with a strong La Niña event in 1999/2000. There were also significantly elevated H$^+$, N, and Mn depositions during the annual biomass burning period in the Amazon basin. Elevated H$^+$ deposition during the biomass burning period caused elevated base metal loss from the canopy and the organic horizon and deteriorated already low base metal supply of the vegetation. Nitrogen was only retained during biomass burning but not during non-fire conditions when deposition was much smaller. Therefore biomass burning-related aerosol emissions in Amazonia seem large enough to substantially increase element deposition at the western rim of Amazonia. Particularly the related increase of acid deposition impoverishes already base-metal scarce ecosystems. As biomass burning is most intense during El Niño situations, a shortened ENSO cycle because of global warming likely enhances the acid deposition at my study forest.

(2) Storm events causing near-surface water flow through C- and nutrient-rich topsoil during rainstorms were the major export pathway for C, N, Al, and Mn (contributing >50% to the total export of these elements). Near-surface flow also accounted for one third of total base metal export. This demonstrates that storm-event related near-surface flow markedly affects the cycling of many nutrients in steep tropical montane forests. Changes in the rainfall regime possibly associated with global climate change will therefore also change element export from the study forest.

Element budgets of Andean tropical montane rain forest proved to be markedly affected by long-range transport of Saharan dust, biomass burning-related aerosols, or strong rainfalls during storm events. Thus, increased acid and nutrient deposition and the global climate change probably drive the tropical montane forest to another state with unknown consequences for its functions and biological diversity.

Resumen

La conservación exitosa de los bosques tropicales de montaña, uno de los ecosistemas más amenzados del planeta, requiere el conocimiento detallado de los procesos biogeoquímicos. De particular interés es la respuesta de los ciclos biogeoquímicos de los elementos a las influencias externas, tales como, la deposición de los elementos o los cambios de clima. Consecuentemente, el objetivo general del presente estudio fue contribuir a la mejor comprensión del rol y el funcionamiento del bosque tropical de montaña. Los objetivos específicos fueron determinar: (1) el rol de los aerosoles transportados de larga distancia y sus mecanismos de transporte; y, (2) el rol de los eventos climáticos extremos de corto plazo, sobre el balance de los elementos del bosque tropical andino de montaña.

En tres microcuencas de 8-13 ha, de un bosque tropical montano, de la vertiente oriental externa de la cordillera Oriental en los Andes en el Sur del Ecuador, en un rango de altitud de 1 850 a 2 200 m s.n.m, entre mayo de 1998 y abril de 2003, desde una aproximación integral de microcuenca y con una resolución al menos semanal, se registraron las concentraciones y los flujos de Ca, Mg, Na, K, NO_3-N, NH_4-N, NOD, P, S, COT, Mn, y Al, en la deposición global, la lluvia que a traviesa el dosel, el lixiviado del mantillo, la solución del suelo a 0.15 y 0.3 m de profundidad, y la escorrentía. También, se utilizó la información meteorologica del área de estudio obtenida por los investigadores cooperantes y del Servio Meteorológico de Brasil (INPE); así como, los datos de los sensores remotos de las agencias espaciales de los Estados Unidos (NASA) y de Europa (ESA).

Los resultados obtenidos evidencian que: (1) entre 1998 y 2003, se registró una fuerte variación interanual en la deposición de Ca [4.4-29 kg ha^{-1} a^{-1}], Mg [1.6-12], y K [9.8-30]). Las altas deposiciones de Ca y Mg cambiaron el balance de las microcuencas, de pérdida a retención, lo que sugiere que el Ca y el Mg, adicionalmente disponibles fueron utilizados por el ecosistema. El incremento de la deposición de metales básicos se relacionó con tormentas de polvo del Sahara y un patrón de precipitación de la Amazonía con períodos secos trans-regionales, lo cual favorece el transporte del polvo hasta los Andes. El incremento de la deposición de metales básicos coincidió con un fuerte evento de La Niña en 1999/2000. Se produjeron también deposiciones elevadamente significativas de H^+, N, y Mn durante el período de quema anual de la biomasa en la cuenca Amazónica. La elevada deposición de H^+ durante el período de quema anual de la biomasa causó una elevada pérdida de metales básicos del dosel y del horizonte orgánico y deterioró la de por si baja provisión de metales básicos a la vegetación. El nitrógeno solamente fue retenido durante la quema de la biomasa, pero no en condiciones de ausencia de quema, durante la cual la deposición fue mucho menor. Por lo tanto, las emisiones de aresoles relacionadas con la quema de la biomasa en la Amazonía, parece que son lo sufientemente grandes para incrementar sustancialmente la deposición de los elementos en el borde occidental de la Amazonía. Particularmente, el incremento relacionado a las deposiciones ácidas empobrecen en mayor grado estos ecosistemas pobres en metales basicos. Puesto que la quema de la biomasa es de mayor intensidad durante las situaciones de El Niño, un ciclo más corto del ENSO ocasionado por el calentamiento global, probablemente incrementaría la deposición ácida en el bosque estudiado.

(2) Los eventos de tormenta que ocasionan el flujo del agua cerca de la superficie a través de C - y la parte superior del suelo rica en nutrientes, fueron las principales vías de exportación de C, N, Al, y Mn (contribuyendo con >50% a la exportación total de estos elementos). Al flujo cerca de la superficie también se le atribuye un tercio de total de la exportación de los metales básicos. Esto demuestra que, el flujo cerca de la superfice relacionado con los eventos de tormenta, marcadamente afecta el ciclo de muchos nutrientes en los bosques tropicales de ladera. Los cambios en el régimen de lluvias, posiblemente asociados con el cambio global del clima, concomitantemente también alterarán la exportación de elementos del bosque estudiado.

El balance de los elementos del bosque lluvioso montano tropical Andino, evidenció que está marcadamente afectado por el transporte de largo rango de los polvos del Sahara, los aerosoles relacionados con la quema de la vegetación, o los fuertes aguaceros durante los eventos de tormenta. Consecuentemente, el incremento de la deposición ácida y de nutrientes y el cambio climático global, probablemente conducirán al bosque tropical de montaña a otro estado con consecuencias desconocidas para su funcionalidad y la diversidad biológica.

Zusammenfassung

Der tropische Bergregenwald zählt zu den am stärksten bedrohten Ökosystemen weltweit. Eine Grundvoraussetzung seines Schutzes ist das Verständnis der biogeochemischen Stoffkreisläufe sowie ihrer Reaktion auf Stoffeinträge und Klimawandel. Daher lag das Hauptaugenmerk meiner Studie darauf, zum verbesserten Verständnis der Rolle und der Funktion des tropischen Bergregenwaldes der Anden beizutragen. Im Detail waren meine Ziele die Bestimmung der (1) Bedeutung und Mechanismen des Langstreckentransportes von Aerosolen und (2) die Auswirkungen kurzzeitiger klimatischer Extremereignisse auf den Stoffhaushalt von Wäldern der tropischen Anden.

In der Zeit von Mai 1998 bis April 2003 bestimmte ich in einem Wassereinzugsgebiets-Ansatz Konzentrationen und Flüsse von Ca, Mg, Na, K, NO_3-N, NH_4-N, DON, P, S, TOC, Mn, und Al im Freilandniederschlag, der Kronentraufe, und dem Bodenwasser unter der organischen Auflage, in 0.15 und 0.3 m Tiefe im Mineralboden sowie im Oberflächenabfluss in mindestens wöchentlicher Auflösung. Die Versuchsflächen waren drei 8-13 ha große Wassereinzugsgebiete, die mit tropischem Bergwald bewachsen und in einer Höhe von 1850-2200m auf der östlichen Abdachung der Ostkordillere in den südecuadorianischen Anden gelegen sind. Ich verwendete meteorologische Daten meines Arbeitsgebietes, die von kooperierenden Wissenschaftlern erhoben wurden, sowie Wetterdaten des brasilianischen Wetterdienstes (INPE). Darüber hinaus benutzte ich Fernerkundungsprodukte der nordamerikanischen und der europäischen Raumfahrtagenturen (NASA und ESA).

Meine Ergebnisse zeigen (1) eine starke interannuelle Variation des Eintrags von Ca [4.4-29 kg ha^{-1} a^{-1}], Mg [1.6-12] und K [9.8-30] in den Jahren 1998-2003. Ein hoher Eintrag führte dabei zu einer Rückhaltung von Ca und Mg im Bestand, während unter Normalbedingungen Ca und Mg aus den Einzugsgebieten ausgewaschen wurden. Dieser Umstand weist darauf hin, dass das zusätzlich eingetragene Ca und Mg durch den tropischen Bergregenwald genutzt wurde. Vermehrter Eintrag von Ca und Mg war dabei mit Sandstürmen in der Sahara und einer Niederschlagsverteilung mit längeren trockenen Abschnitten über Amazonien verbunden, die einen Aerosoltransport aus der Sahara ermöglichten. Dabei fiel der erhöhte Erdalkalielement-Eintrag im hydrologischen Jahr 1999/2000 mit einer ausgeprägten La Niña-Phase des ENSO-Zyklus' zusammen. Des Weiteren beobachtete ich erhöhte H^+, N und Mn-Einträge während Perioden hoher Brandaktivität im Amazonasbecken. Erhöhter Protoneneintrag während der Brandsaison führte dabei zu erhöhter Auswaschung von Erdalkali-Kationen aus dem Kronenraum und der organischen Auflage und verschlechterte so die bereits mangelnde Verfügbarkeit an Erdalkalielementen für die Vegetation weiter. Stickstoff hingegen wurde nur während der Brandsaison zurückgehalten, jedoch nicht während der niedrigeren Einträge außerhalb der Brandsaison. Daraus folgere ich, dass die mit der Biomasseverbrennung in Amazonien verbundenen Emissionen die Stoffeinträge am westlichen Rand des Amazonasbeckens merklich erhöhen. Insbesondere der damit verbundene Anstieg des Protoneneintrages führt zu Verlusten von Erdalkali-Elementen aus dem ohnehin schon erdalkaliarmen Ökosystem. Da die Brandsaison in Amazonien während der El Niño-Phase des ENSO-Zyklus' besonders intensiv ausfällt, verstärkt die fortschreitende globale Erwärmung den sauren Regen in meinem Untersuchungsgebiet.

(2) Der wichtigste Austragsweg von C, N, Al und Mn war durch Sturmereignisse verursachter oberflächennaher Abfluss durch den nährstoffreichen Oberboden, der für bis zu >50% des Gesamtaustrages dieser Elemente verantwortlich war. Der

oberflächennahe Abfluss verursachte auch ein Drittel des Gesamtaustrages an Erdalkalielementen. Dies zeigt, dass die im steilen tropischen Bergregenwald durch Sturmereignisse hervorgerufenen oberflächennahen Wasserflüsse maßgeblich in den Stoffhaushalt zahlreicher Elemente eingreifen. Die mit dem globalen Klimawandel verbundene vermutliche Erhöhung der Niederschlagsintensität könnte also den Nährstoffaustrag meines Versuchsgebietes ändern.

Der Stoffhaushalt andiner Bergregenwälder wurde sowohl durch den Langstreckentransport von Saharastaub und brandbürtigen Aerosolen als auch von starken Regenfällen während Sturmereignissen beeinflusst. Die Protonen- und Nährstoffeinträge in den tropischen Bergregenwald werden durch die fortschreitende globale Erwärmung möglicherweise weiter erhöht. Dies könnte Folgen für Funktion und Biodiversität des Bergregenwaldes haben.

Acknowledgments

I thank Prof. Dr. Wolfgang Wilcke for trusting in my ability to run this project, for encouraging my work, and for the constructive discussions on my findings. Thanks to Prof. Dr. Wolfgang Zech for his support and the opportunity to work in an excellent group at the Soil Science laboratory of the University of Bayreuth. I also thank Prof. Dr. Martin Kaupenjohann, Prof. Dr. Gerd Wessolek and their working groups for the constructive discussions and the good time I spent with them at the Institute of Ecology of the Berlin University of Technology. In addition, I thank my colleagues at the Geographic Institute of the Johannes Gutenberg University Mainz, where I spent the last part of my research odyssey.

I thank Carlos Valarezo and the National University of Loja for the important and persistent support in Ecuador, Dr. Paul Emck for providing meteorological data and Dr. Rütger Rollenbeck for the support in evaluating the remote sensing data, Syafrimen Yasin for the soil and the flux data of the first year and my dear colleagues Dr. Katrin Fleischbein, Dr. Rainer Goller, and Myra Sequeira for providing surface flow and chemical concentration data, as well as many German and Ecuadorian student helpers for their support in data acquisition. Also I thank Prof. Dr. Meinrat Andreae, Prof. Dr. Steven Foley, Dr.Andreas Richter, and Prof. Dr. Heini Wernli for discussion, the NASA for providing the TOMS data and graphs, the NOAA for admission of HYSPLIT data, INPE-CPTEC for the meteorological data of Amazonia and A. Richter for providing the GOME data.

My work was conducted in the frame of the Research Unit "Functionality in a Tropical Mountain Rainforest: Diversity, Dynamic Processes and Utilization Potentials under Ecosystem Perspectives". I thank all the cooperating working groups for providing a stimulating interdisciplinary climate at the research station Estación Científica San Francisco. I am grateful to Nature and Culture International (NCI) in Loja, Ecuador, for providing the study area and the research station and to the Ministerio del Ambiente of the Republic of Ecuador for the permission to conduct this study. My study was funded by the Deutsche Forschungsgemeinschaft (FOR402, Wi1601/5-2, -3).

A Summarizing overview

1. Introduction

Tropical montane rain forests are among the plant and animal richest of the world and now threatened to vanish [*Brooks et al.*, 2002]. To develop long-term conservational strategies and sustainable land-use practices, knowledge of the ecosystem functioning is crucial. This requires understanding the controls of element cycling and the response of element cycles to long-term variations in environmental constraints e.g., rainfall regime, atmospheric inputs, or flow conditions in soil.

In ecosystems in which nutrient demand of the vegetation is not covered by weathering of the parent rock, atmospheric inputs are an important source of plant nutrients, and might help to overcome nutrient limitations [*Hedin et al.*, 2003]. One important source of atmospheric input to Amazonian forests is the Sahara desert [*Kaufman et al.*, 2005; *Reichholf*, 1986; *Swap et al.*, 1992]. The Sahara produces 400-700 Tg of atmospheric dust per year, which represents almost 50% of the global dust production [*Prospero and Lamb*, 2003; *Schütz et al.*, 1981]. A large portion of this dust is transported across the North Atlantic by the predominant westerly winds (240±80 Tg [*Kaufman et al.*, 2005]).

It has been hypothesized but never directly shown that desert dust helps reducing nutrient limitations in the Amazonian rain forests [*Okin et al.*, 2004; *Reichholf*, 1986; *Swap et al.*, 1992]. Among the elements considered to be potentially fertilizing are base metals (i.e., K, Ca and Mg), which can be abundant in Saharan dust [*Claquin et al.*, 1999; *Moreno et al.*, 2006; *Reid et al.*, 2003]. Base metal supply is particularly low in Amazonian ecosystems because of highly weathered soils and the absence of local base metal sources [*Cuevas and Medina*, 1986; *Swap et al.*, 1992; *Wilcke et al.*, 2001].

It is, however, unclear if Saharan dust can be transported in significant quantities across the usually humid Amazon basin towards the remoter parts of Amazonia and the montane forests of the Andes. As aerosols are almost instantly scavenged by rain, the Amazon basin should act as a wet barrier for Sahara dust transport. Therefore, Saharan dust can only reach remote Amazonian forests at the outer rim of the Amazon basin during dry spells allowing for dust transport across the region. Variations in the hydrological regime of Amazonia have often been linked to the El Niño Southern Osciliation [ENSO; e.g. *McPhaden*, 1999].

Another major source of aerosols to the atmosphere is biomass burning [*Andreae et al.*, 1988]. These aerosols can contain high concentrations of C, plant nutrients (particularly N, S, P, and K), and acids [*Allen and Miguel*, 1995; *Artaxo et al.*, 2002; *Da*

Rocha et al., 2005]. Some of the aerosol species can be transported over large distances in the atmosphere [*Pereira et al.*, 1996] and deposited to distant ecosystems [*Mahowald et al.*, 2005]. Amazonia is a source region of aerosols for downwind ecosystems such as the Andean tropical montane forest in Ecuador because of forest destruction by fire and growing agricultural activity in the region [*Kauffman et al.*, 1995]. To explore the effect of Amazonian fire-derived deposition in the Andes, direct monitoring of wet and dry deposition is required.

To estimate the effect of elevated element deposition to forest ecosystems, budgets of the various ecosystem strata (e.g., canopy budget or organic horizon budget [*Clark et al.*, 1998]) and the whole ecosystem can be used [*Bruijnzeel*, 1991]. Comparison of separate budgets of these strata between burning and non-burning periods offer specific insight into the effects of biomass burning in Amazonia on the remote montane ecosystems of the Andes. The complete nutrient budget of a forest ecosystem requires a closed budgeting unit and consideration of all inputs and outputs. Unfortunately determining nutrient export is complicated by the great variability of element concentrations in discharge waters [*Godsey et al.*, 2004; *McDowell and Asbury*, 1994].

In many studies, positive correlations between the discharge rate and element concentrations were observed e.g., for dissolved organic carbon [DOC, *Buffam et al.*, 2001; *Hook and Yeakley*, 2005] or total dissolved nitrogen [TDN, *Campbell et al.*, 2000; *Goller et al.*, 2006]. In preceding studies at my site in southern Ecuador, increasing concentrations of DOC, DON, NO_3-N, and partly of NH_4-N in stream water were reported when discharge was high. The concentrations of these chemical constituents were greatest during periods of heavy rain [*Goller et al.*, 2006; *Wilcke et al.*, 2001].

Previous work at my study site by means of a $\delta^{18}O$ approach showed that a large portion of water flow in soil during rainstorms indeed occurred laterally, whereas during baseflow conditions most of the stream water originated from the deeper mineral soil [*Goller et al.*, 2005]. Associated with the finding that concentrations of DOC, NO_3-N, NH_4-N, DON, S, and P at my study site are greatest in the organic soil layers [*Wilcke et al.*, 2002], this suggested a relation of element export to the water flow regime and particularly to the depth of water flow in the soil.

The concentrations of organic compounds and organically bound nutrients in stream water are likely to be particularly influenced by lateral flow, because soluble organic matter accumulates in the organic layer and topsoil. This accumulation of organic matter also influences metal mobility in soil, as it is closely related to pH and the concentrations of ligands forming soluble metal complexes. Particularly important ligands are contained in the dissolved organic matter [DOC, *Rieuwerts*, 2007].

Studies combining hydrological and biogeochemical approaches to elucidate processes of stormflow-related element export are scarce, particularly in the Tropics [*Saunders et al.*, 2006]. Especially in steep, forested catchments responding quickly to storm events (within minutes or hours), stream chemistry is much more influenced by changing environmental conditions than would be the case in slower reacting systems of lesser slope [*Schellekens et al.*, 2004]. This further complicates the extrapolation of single storm event data collected at high resolution to longer periods as is done when using the "storm chasing" approach [*Buffam et al.*, 2001] or flow separation approaches, which have proven suitable for catchments of lesser slope and/or greater size [e.g., *Evans et al.*, 2004]. In contrast, studies sampling at fixed intervals over longer periods, as is done in long-term ecosystem studies [e.g., *Matzner*, 2004], usually calculate element export by multiplying flow-weighted mean concentrations with cumulative flow and face difficulties in assessing the influence of storm events on element export. This problem might be solved by defining a classification key based on discharge rate in order to estimate the contribution of different flow regimes to element export for long time series.

By monitoring all ecosystem fluxes for five consecutive years between 1998 and 2003 in weekly resolution using a whole-catchment approach in an Andean lower tropical montane rain forest in south Ecuador I tested the following hypotheses:

(i) Saharan dust transport adds significant amounts of base metals to montane forests at the western rim of the Amazon basin

(ii) the nutrient deposition via Saharan dust influences the catchment budget of base metals in an Ecuadorian montane forest

(iii) dust transport across Amazonia is related to the precipitation pattern controlled by El Niño Southern Oscillation (ENSO) and therefore is an overseasonal phenomenon.

(iv) biomass burning emissions of Amazonia and other sources upwind of the north-eastern trade winds are transported to and deposited at tropical montane forests in the Andes.

(v) Biomass burning-related deposition contributes to forest nutrition, as indicated by the retention of deposited elements in the ecosystem.

(vi) the concentrations of chemical constituents in stream water of steep, forested catchments are related to discharge levels in a way that is specific for each chemical constituent.
(vii) the depth of water flow in soil determines the concentrations of chemical constituents in stream water.
(viii) discharge level classification is a suitable tool to estimate the contribution of different flow regimes to element export from steep, forested catchments for long time series.
(ix) storm events have a significant influence on catchment nutrient export.

2. Material and Methods

2.1 Study site

The study area is located on the eastern slope of the "Cordillera Real", the eastern Andean cordillera in south Ecuador facing the Amazon basin at 4° 00` S and 79° 05` W. I used three 30-50° steep and 8-13 ha large microcatchments (MC1-3) under montane forest at an altitude of 1900-2200 m above sea level (a.s.l.) for my study (Figure A-1). The equipment was installed in each MC on transects, about 20 m long with an altitude range of 10 m, on the lower part of the slope at 1900-1910 m a.s.l. (transects MC1, MC2.1, and MC3). Moreover, extra instrumentation was installed at 1950-1960 (MC2.2) and 2000-2010 m a.s.l. (MC2.3). All transects were located below closed forest canopy and aligned downhill. Three deforested sites near these microcatchments were used for rainfall gauging. Gauging site 2 existed since April 1998, gauging sites 1 and 3 were built in May 2000. All catchments drain via small tributaries into the Rio San Francisco which flows into the Amazon basin.

Within the monitored period between April 1998 and April 2003 annual precipitation ranged between 2340 and 2667 mm. Additional climate data were available from a meteorological station [*Richter*, 2003] between MC 2 and 3 (Figure A-1). June tended to be the wettest month with 302 mm of precipitation on average, in contrast to 78 mm in each of November and January, the driest months. The mean temperature at 1950 m a.s.l. was 15.5 °C. The coldest month was July, with a mean temperature of 14.5 °C, the warmest November with a mean temperature of 16.6 °C.

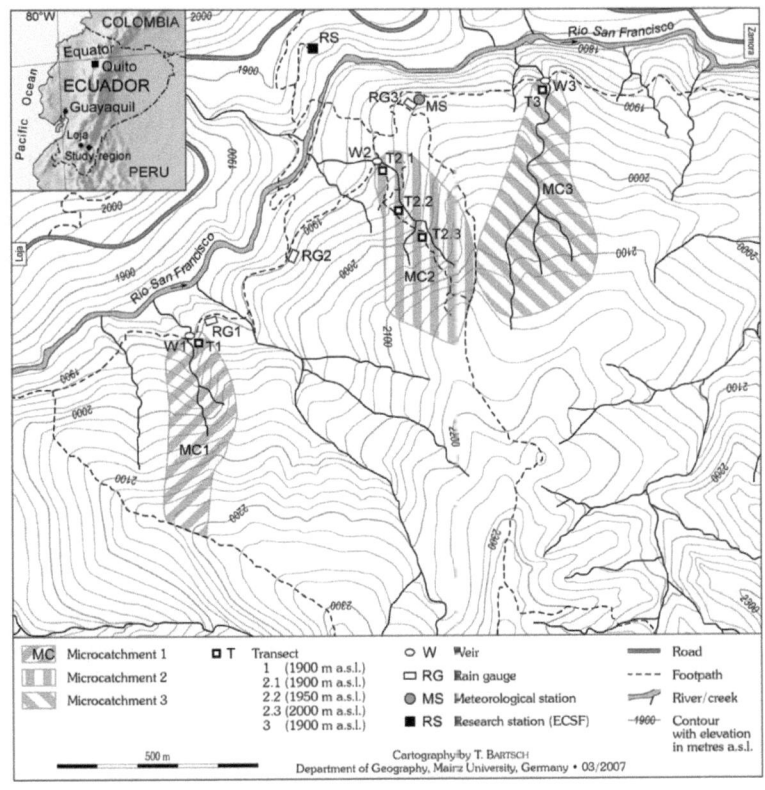

Figure A-1: Location of the study area

Recent soils have developed on postglacial landslides or possibly from periglacial cover beds [*Wilcke et al.*, 2001; 2003]. Soils are Humic Eutrudepts on transect MC1, Humic Dystrudepts on transects MC2.1, MC2.2, and MC2.3, and Oxyaquic Eutrudepts on transect MC3 [*USDA-NRCS*, 1998]. All soils are shallow, loamy-skeletal with high mica contents. The organic layer consisted of Oi, Oe, and frequently also Oa horizons and had a thickness between 2 and 43 cm [mean of 16 cm; *Wilcke et al.*, 2002]. The thickness increased with increasing altitude giving Histosols (mainly Terric Haplosaprists) above c. 2100 m. Selected soil properties are summarized in Table A-1; data were taken from the work of *Yasin* (2001). The underlying bedrock consists of interbedding of paleozoic phyllites, quartzites and metasandstones (the "Chiguinda unit" of the "Zamora series" in the work of *Hungerbühler* [1997]).

Table A-1: Means and ranges (in brackets) of selected soil properties in an Ecuadorian lower montane forest (n=47).

	Units	Horizon		
		O	A	B
ECEC[a]	[mmol$_c$ kg^{-1}]	NA[b]	72 (15-152)	57 (13-135)
BS[c]	[%]	NA[b]	27 (1.4-95)	24 (4.4-100)
pH	(H$_2$O)	4.0 (3.2-6.7)	4.3 (3.7-6.4)	4.8 (3.6-6.3)
C	[g kg^{-1}]	395 (244-503)	34 (2.8-92)	19 (2.6-50)
N	[g kg^{-1}]	23 (15-32)	3.4 (0.52-10)	2.1 (0.55-5.1)
P	[g kg^{-1}]	0.96 (0.42-2.1)	0.99 (ND[d]-4.4)	0.90 (ND[d]-4.7)
S	[g kg^{-1}]	2.8 (1.8-8.9)	0.38 (0.12-0.92)	0.31 (0.11-1.8)
Ca$_{tot}$[e]	[g kg^{-1}]	6.2 (0.31-19)	0.51 (0.02-2.9)	0.49 (0.09-2.8)
Ca$_{exch}$[f]	[mmol$_c$ kg^{-1}]	NA[b]	9.5 (0.10-63)	2.5 (0.20-11)
Mg$_{tot}$[e]	[g kg^{-1}]	2.0 (0.25-5.8)	0.76 (0.05-1.6)	0.75 (0.24-1.5)
Mg$_{exch}$[f]	[mmol$_c$ kg^{-1}]	NA[b]	11 (0.24-74)	3.6 (0.05-16)
K$_{tot}$[e]	[g kg^{-1}]	3.4 (0.63-9.5)	16 (0.38-28)	17 (5.3-32)
K$_{exch}$[f]	[mmol$_c$ kg^{-1}]	NA[b]	1.6 (0.41-4.0)	0.80 (0.25-2.3)
Na$_{tot}$[e]	[g kg^{-1}]	0.11 (0.03-0.33)	1.3 (0.37-4.1)	1.3 (0.40-3.8)
Na$_{exch}$[f]	[mmol$_c$ kg^{-1}]	NA[b]	2.7 (0.30-3.7)	2.9 (0.30-6.1)
Al$_{tot}$[e]	[g kg^{-1}]	7.8 (1.2-32)	47 (5.2-84)	52 (34-80)
Al$_{exch}$[f]	[mmol$_c$ kg^{-1}]	NA[b]	46 (1.2-131)	47 (0.06-128)
Mn$_{tot}$[e]	[g kg^{-1}]	0.35 (0.02-2.6)	0.46 (0.14-1.0)	0.42 (0.17-1.2)
Mn$_{exch}$[f]	[mmol$_c$ kg^{-1}]	NA[b]	0.19 (ND[d]-2.7)	0.28 (ND[d]-1.5)

[a]Effective cation-exchange capacity
[b]NA, not analysed.
[c]Base saturation.
[d]ND, not detected.
[e]The suffix "tot" indicates the total concentration of an element.
[f]The suffix "exch" indicates the exchangeable concentration of an element.

MCs 2 and 3 are entirely forested, whereas the upper part of MC 1 has been used for agriculture until about 10 years ago. This part is currently undergoing natural succession and is covered by grass and shrubs. The study forest can be classified as "bosque siempreverde montaño" (evergreen montane forest [*Balslev and Ollgaard*, 2002]) or as Lower Montane Forest [*Bruijnzeel and Hamilton*, 2000]. More information on the composition of the forest can be found in the work of *Homeier* [2004].

2.2 Field sampling

Water samples were collected between April 1998 and April 2003. Each gauging station for incident precipitation consisted of five samplers. Solution sampled by rainfall collectors was "bulk precipitation" [*Whitehead and Feth*, 1964], since collectors were open to dry deposition between rainfall events [*Parker*, 1983]. However, the contribution of dry deposition to my rainfall collectors only comprised the soluble part of particles that are large enough to sediment gravitationally. I assumed that this coarse particulate deposition was small compared with the far larger aerosol trapping capacity of the forest canopy collecting also gaseous and fine particulate deposition by impaction and therefore neglected the coarse particulate deposition [*Parker*, 1983]. Each of the five transects was equipped with five throughfall collectors (in May 2000, three more collectors were added on each transect). All throughfall samplers had a fixed position that was arbitrarily chosen and evenly distributed along the transects. To rove samplers after each sample collection, as suggested by *Lloyd & Marques* [1988], to improve the representativity of the sample would have resulted in an unacceptable damage to the study forest that was only accessible on very steep machete-cleared and rope-secured paths. More information on the throughfall measurement is reported in the work of *Fleischbein et al.* [2005].

At the lowermost transects of all catchments (MC1, MC2.1, MC3) five trees were equipped with stemflow collectors. Each of the throughfall and stemflow samples were combined to one sample per transect in the field. Stream water samples were weekly taken from the center of the streams at the outlet of each catchment.

Throughfall and rainfall collectors consisted of fixed 1-l polyethylene sampling bottles and circular funnels with a diameter of 115 mm. The opening of the funnel was at 0.3 m height above the soil. The collectors were equipped with table tennis balls to reduce evaporation. Incident rainfall collectors were additionally wrapped with aluminum foil to reduce the impact of radiation. Stemflow collectors were made of polyurethane foam and connected with plastic tubes to a 10-l container [*Likens and Eaton*, 1970]. In each catchment, four trees of the uppermost canopy layer and one tree fern belonging to the second tree layer were used for stemflow measurements. The species were selected to be representative of the study forest although this was difficult because of its high plant diversity. A list of the selected species is given by *Fleischbein et al.* [2005]. Litter leachate was sampled by zero tension lysimeters, consisting of plastic boxes (0.20 x 0.14 m sampling area) covered with a polyethylene net (0.5 mm mesh width). The boxes were connected to 1-litre polyethylene sampling bottles with a plastic tube. The lysimeters were installed from a soil pit below the organic layer and parallel to the surface. The organic layer was not disturbed; most roots in the organic layer remained intact [*Wilcke et al.*, 2001]. Mineral soil solution was sampled by suction lysimeters (mullite suction cups, 1 µm ± 0.1 µm pore size) with a vacuum pump. The vacuum was held permanently and adjusted to the matric potential. The lysimeters did not collect the soil solution quantitatively.

I used the results of element analyses of the Oi, Oe, Oa, A, and B horizons of 47 soils, distributed so as to represent the soils proportionally all three catchments (Wilcke et al., 2002).

2.3 Hydrological measurements

Rainfall, throughfall and stemflow were measured weekly by recording single volumes for each collector. Additionally each catchment was equipped with a tipping bucket rain gauge (NovaLynx® 260-2500, NovaLynx Corporation, Grass Valley, U.S.A.) to obtain higher resolution data of throughfall volume. Due to logger breakdowns and funnel blockings the data set was incomplete. Missing data was substituted by regression of precipitation data (RF) of the meteorological station on throughfall (TF, shown in Section A, p. 45)

Water fluxes in soil were modeled by modifying the Soil Water Balance (SWB) model [*DVWK (Deutscher Verband für Wasserwirtschaft und Kulturbau)*, 1996]. The SWB determines water fluxes out of predefined soil layers as the sum of throughfall and stemflow volume (input) minus independently determined transpiration (output) minus (or plus) change in stored water in the soil layers as calculated by the difference in water contents of the respective soil layer between two soil water measurements. For deeper soil layers, the input is the output of the overlying soil layer. I assumed direct evaporation from the soil as negligible and derived weekly transpiration rates by partitioning the annual difference between throughfall and discharge of each catchment proportionally to the weekly evapotranspiration rates as modeled by REF-ET (described in *Fleischbein et al.* [2006]). Weekly transpiration rates were furthermore split between the soil layers according to the root length densities of the respective soil layer taken from *Soethe et al.* [2006], assuming a linear relationship between water uptake of the vegetation and fine root abundance. I used soil water-content measurements logged by FDR probes in transect MC 2.1 at 0.1, 0.2, 0.3, and 0.4 m depths for all transects since differences in soil water content were little pronounced because of the overall wet environment of the study site [*Fleischbein et al.*, 2006]. Data gaps of soil water fluxes (because of lacking soil water contents) were substituted with the help of a regression model of weekly soil water fluxes on weekly throughfall volumes. This approach is biased because I assumed that all water flow in soil was vertical which was not the case. However, there is no quantification of lateral flow available.

To quantify surface flow, in April 1998 Thompson (V-notch) weirs (90°) with sediment basins were installed in the lower part of each catchment and water level was instantaneously recorded hourly with a pressure gauge (water level sensor). Additionally, water level was measured by hand after sampling of stream water. The empirical equations shown in Section A, p. 45 were used.

Unfortunately, logger breakdowns occurred during the runoff measurement likely because of the frequently wet conditions in the studied forest. Data gaps were closed by means of the hydrological modelling program TOPMODEL [*Beven et al.*, 1995] as described by *Fleischbein et al.* [2006].

2.4 Chemical analyses

In the solution samples, Cl^- concentrations were determined with a Cl^--specific ion electrode (Orion® 9617 BN) immediately after collection in Ecuador during the first three years. In the fourth and fifth year, Cl^- was analyzed with a segmented continuous flow analyzer (CFA San plus, Skalar®). After export of the filtered 100-ml aliquots from Ecuador to Germany in frozen state, Ca, Mg, K, and Na concentrations were determined with flame atomic absorption spectroscopy (AAS). Ca, Mg, K, and Na are referred to as "base metals" in the following. Na and Cl^- concentrations are used to assess the potential marine influence on the chemical composition of rain water. Furthermore, water samples were analyzed colorimetrically with a Continuous Flow Analyzer (CFA) for concentrations of dissolved inorganic nitrogen (NH_4-N and NO_3-N + NO_2-N, hereafter referred to as NO_3-N), and total dissolved nitrogen (N_{tot}, after UV oxidation to NO_3). Additionally, total dissolved sulfur and total dissolved phosphorus concentrations were determined by Inductively-coupled Plasma Optical Emission Spectrometry (ICP-OES, Integra XMP, GBC Scientific Equipment®). Total organic carbon (TOC) concentrations were determined with an automatic TOC analyzer (TOC 5050, Shimadzu®) and Al and Mn concentrations via Inductively-coupled Plasma-Mass Spectrometry (ICP-MS, VG PlasmaQuad PG2 Turbo Plus, VG Elemental®).

Some samples had concentrations below the detection limit of the analytical methods (0.075 mg l^{-1} for N, 0.2 mg l^{-1} for P, 0.3 mg l^{-1} for S, 0.001 mg l^{-1} for Ca, Mg, K, and Na, 0.005 mg l^{-1} for Al, and 0.002 mg l^{-1} for Mn). For calculation purposes, values below the detection limit were set to zero (for Ca, Mg, Na, K, and TOC: <0.01%, N and Zn <1%, Al <7%, S <25%, Mn <30%, and P <45% of the values were below detection limit). Thus, my annual means underestimate the real concentration of chemical constituents and mean concentrations can be smaller than the detection limit.

2.5 Remote sensing and transport pathway reconstruction

To compare tropospheric dust emission of the Sahara to atmospheric input to Andean montane rainforest I used the aerosol index of the "NASA Ozone Processing Team" based on measured radiances recorded by the Total Ozone Mapping Spectrometer (TOMS [*Torres et al.*, 2002]) on board the NASA satellite "EARTH Probe". I used weekly and monthly averages of daily TOMS aerosol index (AI) data. For reconstruction of transport pathways of air to my study site I calculated 13-day backward trajectories (target heights: 3000, 4000, and 6000m. a. s. l., all under the average cloud top heights at the study site [*Bendix et al.*, 2006]) using the Hybrid Single Particle Lagrangian Integrated Trajectories (HYSPLIT) model [*Draxler and Hess*, 1998] provided by the NOAA Air Recources Laboratory (ARL). To validate backward trajectories crossing the intertropical convergence zone I additionally calculated 13 day forward trajectories from western African sources (30°N, 9°W; 23°N, 15°W; 15°N, 16°W).

To assess dust transport conditions across Amazonia I evaluated a typical Saharan dust transport corridor representing >85% of noon trajectory passage at 3000, 4000, and

6000 m target heights across Amazonia by backward trajectory analysis (HYSPLIT). This corridor was split into three zones, each roughly representing the distance across which Saharan dust can be transported in a single day. I calculated a daily mean precipitation based on the data of 5-12 meteorological stations in each zone run by the Instituto Nacional de Pesquisas Espaciais ([INPE]; ~300,000 measurements, data with obvious errors [<1%] were removed, data are available at http://tempo.cptec.inpe.br:9080/PCD). I considered the transport conditions for dust as optimal, if Saharan air was able to cross a zone without being exposed to precipitation (=0 mm mean daily precipitation of all meteorological stations in the considered zone). Furthermore, I assumed that this had to be the case on three consecutive days from zone 1 in the east to zone 3 in the west. I took the number of such days per month (counted as date of the trajectory's arrival at my study site) as an index of the transport probability of Saharan dust over Amazonia.

To compare biomass burning seasons with periods not affected by biomass burning I assigned each sampling week to one of the two classes "fire" (biomass burning) and "no fire". Therefore, I used 13-day backward trajectories combined with a fire pixel dataset by counting fire pixels on the pathway of each trajectory in a radius of 25 km when monitored with a tolerance of one day backwards around the actual parcel location. A pixel was considered to contain fire, if the relevant temperature channel of the sensor detected average temperatures above $70°C$. The firepixel data was obtained from the satellites NOAA-12 for the whole study period. To improve the accuracy of my fire pixel count I additionally used data of the satellites GOES-8 (Geostationary Environmental Satellite) and MODIS 01D (Moderate Resolution Imaging Spectroradiometer) for the year 2002. A fire pixel was considered as identified, if it was classified as such by at least one of these three satellites. The fire pixel data are provided by the Brazilian Space Agency INPE (Instituto Nacional de Pesquisas Espaciais; http://sigma.cptec.inpe.br/produto/queimadas). More details of the fire pixel count are given in the work of Rollenbeck et al. [2008].

Afterwards I tested the consistency of the firepixel data with the GOME (Global Ozone Monitoring Experiment) NO_2 remote sensing product [*Richter et al.*, 2005] for $4°S$ $79°W ± 2°$ (= the air column over the study site). Biomass burning periods were defined as the sum of all weeks when the GOME NO_2 index was higher than the 5-year mean and more than 14 fire pixels (25%-percentile) were counted on the pathway of this week's trajectories.

2.7 Calculations and statistical analyses

Annual element fluxes were calculated for rainfall, throughfall, and stemflow by multiplying the respective annual volume-weighted mean (VWM) concentrations with the annual water fluxes. Element fluxes with surface runoff were calculated by multiplying flow-weighted mean (FWM) concentrations with the measured or modeled annual surface runoff and referring the annual flux to the surface area of the catchments (MC1: 8 ha, MC2: 9.1 ha, MC3: 13 ha). To estimate the dry deposition and quantify canopy leaching I used the model of *Ulrich* [1983] as described in detail in Section A, pp. 46-47.

To test whether the interannual variation in base metal deposition was related to biomass burning I performed a principal component analysis after varimax rotation (SPSS® 13.0 for windows®) for Ca, Mg, K, N species, and S. If elements loaded the same principal component I considered them to have the same or collocated sources.

For description of the water flow regime, stream water samples collected in weekly interval at my weirs were grouped into five flow classes representing the type of discharge event at the time of sampling. Flow classes were defined by their relation to the 5-year mean discharge for each catchment as modelled with TOPMODEL [*Fleischbein et al.*, 2006]. Discharge of flow class "superdry" was defined as less than 25%, flow class "baseflow" as between 25% and 50%, and flow class "intermediate" as ranging between 50% and 200% of the 5-year mean discharge of a catchment.

High discharge periods were further divided into the flow classes "stormflow" and "lateral flow". Stormflow was defined as occurring if the discharge was more than double of the 5-year mean. Lateral flow met the same criterion but in addition was associated with an at least four times greater 12-hour throughfall than the 5-year mean. The throughfall criterion substitutes the soil water saturation criterion of >85% of the maximum soil water content of the O horizon and the uppermost 0.2 m of the mineral soil (but not necessarily for the 0.2-0.4 m depth layer). This substitution was necessary because of frequent breakdowns of the FDR data logger in MC2 and no available FDR data in MC1 and MC3. The throughfall criterion proved to be equally valid in 100% of the lateral flow events where both throughfall and soil water content data were available (>50% of all events in MC2).

Based on the assumption that the flow class "superdry" with its lowest discharge rate represents the concentrations of chemical constituents in groundwater, I calculated a "flow class effect" (FE) to estimate the contribution of each flow class to the concentrations of chemical constituents in stream water (see Section D, p. 97).

To estimate the contribution of different flow classes to total element export I developed a simple new model, in which I classified hourly discharge [obtained with TOPMODEL, *Fleischbein et al.*, 2006] into the five flow classes. Then, I multiplied the cumulative discharge for a given flow class during the monitored 5-year period with the mean concentration of each of the chemical constituents studied in the same flow class. Finally, the export rates of each flow class and each chemical constituent were summed to give a total 5-year element export for each catchment. To validate this new approach, I compared its results with conventionally derived results from flow-weighted means and cumulative discharge.

To further investigate into the controls of metal solubility during the various flow conditions, I calculated humic complexation of metals in water samples by the speciation model VisualMinteQ v. 2.53 [*Gustafsson et al.*, 2001]. I used the Nica-Donnan model sub-routine with the active DOM to DOC ratio set to 1.4 and assuming 100% of active DOM to be fulvic acids. Furthermore, I assumed concentrations of TOC to be equal to concentrations of DOC although my samples were filtered through 4-7 µm pores. I tested the stability of the observed trends in metal complexation over the five flow classes by assuming that only 50% of TOC was DOC and did not observe significant changes.

I selected two key target variables (Ca and N_{tot}) to represent the base metal and the nitrogen species cluster formed by the principal component analysis (with varimax rotation) in order to reduce number of target variables (Figure A2) in a general linear model (GLM, type I sum of squares; Na, and Cl were not tested because of their low importance for plant nutrition at the study site). For the concentrations of each of Ca and N_{tot} in all ecosystem fluxes, a separate GLM was calculated with a set of explanatory variables fitted as covariables in the GLM (listed below). In all models, I fitted the site effect (i.e. the three studied catchments) first to remove the influence of the specific catchment properties from the further analysis. Because the order in which variables were fitted matters, I compared the results of different hierarchical models. For each hierarchical step of the GLM all explanatory variables were tested for their ability to explain the variance of the target variable. Afterwards the variable which explained the variance of the target variable best was chosen for the tested hierarchical level of the GLM, irrespectively of the significance of this variable. In a next step all remaining explanatory variables were tested newly in the described manner on the hierarchically lower second level and so on until all explanatory variables where placed in hierarchic order concerning their explanatory power of the variance of the target variable. Once this stage was completed all possible interactions of explanatory variables were tested, again listed in the order of their explanatory power for variance. The analyses were conducted with a commercial software package (SPSS 15, SPSS Inc. Chicago, IL, USA).

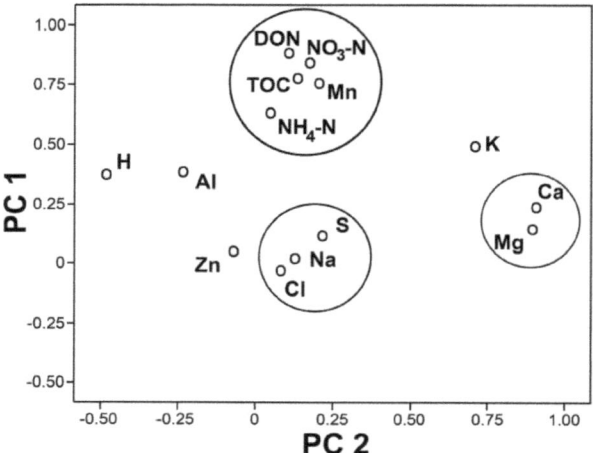

Figure A2: Illustration of the factor loading of element concentrations in rainfall between May 1998 and April 2003 in the coordinate system of principal components (PC) 1 and 2. Circles indicate element grouping.

The set of explanatory variables consisted of:

Firepix: the under "Long range trajs" described trajectories (but without the criterion of source region Africa) combined with a fire pixel dataset by counting fire pixels on the pathway of each trajectory in a radius of 25 km when monitored with a tolerance of one day backwards around the actual parcel location.

GOME: NO_2 remote sensing product of the Global Ozone Monitoring Experiment [GOME, *Richter et al.*, 2005] for 4°S 79°W ± 2° (= the air column above the study site)

Humidity: relative humidity of air at the Meteorological Station at the study site (Figure A1).

Long-range trajs: number of 13-day backward trajectories per week originating at least in northern Africa (or further to the east) representing strong trade wind systems (target heights: 3000, 4000, and 6000m. a. s. l., all under the average cloud top heights at the study site.

P h/wk: rainfall duration. Mean of the hours with precipitation per week at the Meteorological Station (Figure A1).

P mm/h: rainfall intensity. Weekly mean of hourly rainfall values [mm].

P transp. corridors: precipitation [mm] in the transport corridor of the trade winds (10°N–10°S, 79°W–40°W, describing the probability of aerosol scavenging by rainfall). Data were obtained of 24 meteorological stations run by the Brazilian Space Agency INPE (Instituto Nacional de Pesquisas Espaciais; http://tempo.cptec.inpe.br:9080/PCD; ~300.000 measurements, data with obvious errors [<1%] were removed).

Radiation: radiation at the study site

Soil water content X: soil water content in the organic horizon ("O_h") or at the 0.2 or 0.4 m depths in the mineral soil ("0.2m" and "0.4m").

SST ENSO 3: sea surface temperature anomalies at the ENSO 3 region [*McPhaden*, 1999].

Stormlateral: number of periods when storm events caused high runoff volumes out of a microcatchment (=stormflow). Stormflow was defined as occurring if the discharge was more than double the 5-year mean.

Temp ECSF: local temperature at the Meteorological Station between MC2 and MC3 (Figure A1).

TOMS: aerosol index of the „NASA Ozone Processing Team" based on measured radiances recorded by the Total Ozone Mapping Spectrometer [TOMS, *Torres et al.*, 2002] on board the NASA satellite "Earth Probe" between 0 to 8°S and 75–80°W.

Tsoil0m & Tsoil0.6m: surface temperature and soil temperature at the study site. Measured inside the stand close to the Meteorological Station (Figure A1).

Windvel: local wind velocity measured at the Meteorological Station (Figure A1). Only represents local wind systems, mainly downslope wind.

X conc Y and X flux Y : concentration or flux inside the ecosystem. "X" is the type of water flow as described above, "Y" is the chemical abbreviation of an element. Example: "TF conc Ca" is the concentration of Calcium in throughfall.

X volume: water volume [mm] of a certain flow in the ecosystem. "RF" = rainfall, "TF" = throughfall (including stemflow), "LL" = litter leachate, "SS15" and "SS30" = soil solution in 0.15m and 0.3m depth of the mineral horizon, "SF" = surface flow.

3. Results and Discussion

3.1 Tropical Andean forest derives calcium and magnesium from Saharan dust (Section B)

I observed a strong interannual variation in Ca deposition. This variation in Ca deposition changed the Ca budget of the investigated catchments from overall loss during low-input years to overall retention during high-input years, indicating a direct response of the ecosystem to Ca deposition. One possible explanation of the retention of a nutrient in the ecosystem is the shortage in supply of the retained nutrient to the vegetation. Retention of Ca was observed at my study sites from 1999 to 2002. About 15 kg of the deposited Ca per hectare were retained in the forest in 1999/2000. The retention of additionally deposited Ca between 1999 and 2002 might indicate that Ca enhanced plant growth, possibly by altering N uptake [*Hedin et al.*, 2003; *McLaughlin and Wimmer*, 1999], and would imply that nutrient accretion because of accelerated plant growth increased.

The temporal course of Mg deposition during the monitoring period strongly resembled that of Ca (as indicated by the close correlation of Ca and Mg concentrations, r=0.92; n=263 weeks). Retention of Mg in the catchments was observed during the hydrological year of 1999/2000 with its high base metal inputs and during the following year (6 and 0.9 kg ha^{-1}, respectively). I interpret the Mg retention in the system as an indirect effect of the assumed altered nutrient accretion induced by Ca deposition.

My results demonstrate that the variation in base metal input by deposition from the atmosphere among different years has a measurable effect on the nutrient budget of a tropical montane rain forest. Given its least availability of all base metals (on a molar basis) and its strongest retention in the studied catchments, I suggest that Ca played the key role for the observed effects of base metal deposition to the monitored forest.

Local dust, anthropogenic emissions, volcanism, biomass burning, and biogenic aerosols could be excluded as a possible explanation for the observed interannual variation in base metal deposition, pointing at long-range transport. Long-range transport of soil dust from the Amazon basin is not a likely source of base metals because of the low alkaline earth metal concentrations of Amazonian ecosystems on predominantly deeply weathered and nutrient-depleted soils [*Cuevas and Medina*, 1986; 1988; *Swap et al.*, 1992]. The only significant atmospheric source of base metals for Amazonia is Saharan dust, which contains appreciable amounts of Ca, Mg, and K (up to 18wt% of Ca oxides, calcites, and dolomites, up to 4wt% of mafic Mg and 3wt% of K_2O [*Claquin et al.*, 1999; *Moreno et al.*, 2006; *Reid et al.*, 2003]). Therefore, I compared the temporal course of aerosol emission activity over the Sahara as detected by TOMS with the temporal course of base metal deposition at my study site. The amount of deposited Ca and Mg corresponded well to Saharan source activity as detected by TOMS. Strong desert storms leading to high aerosol concentrations over the west Sahara, a known source for Ca- and Mg-rich aerosols [*Claquin et al.*, 1999; *Moreno et al.*, 2006; *Reid et al.*, 2003], corresponded to periods of high Ca and Mg deposition at my study site. Source activity in the eastern Sahara or the Sahel, both known to be substantially weaker emission sources for Ca and Mg [*Claquin et al.*, 1999; *Moreno et al.*, 2006; *Reid et al.*, 2003] caused much lower Ca and Mg depositions. When Saharan aerosol emissions were low or absent, I never observed elevated Ca or Mg concentrations in bulk deposition higher than 0.5 mg l^{-1} during the period of elevated base cation deposition.

The number of days with optimal transport conditions for dust in the Amazonian atmosphere was different among the monitored hydrological years. Days with optimal transport conditions mainly occurred during 1999/2000 and implied a difference in rainfall distribution among the years. During "La Niña", strong rain events were interrupted by dry spells allowing for dust transport. During periods of climate conditions not affected by ENSO, precipitation was more equally distributed, although data suggest a slightly elevated "background" cation deposition when strong aerosol emission activity in the Sahara met high numbers of trajectories to the study side (e.g., between January to April 1998, December 2000 to April 2001, and November 2001- May 2002). This might indicate that a small part of desert dust might be able to cross a wet Amazon basin.

More than 95% of weekly samples with elevated Ca and Mg deposition (>0.5 mg l^{-1} Ca; >0.25 mg l^{-1} Mg) in the Ecuadorian forest were collected during periods of optimal transport conditions. This corresponds to the finding that high Saharan dust deposition to Amazonia is following rainfalls of major wet systems [*Swap et al.*, 1992]. Therefore, I hypothesize that Saharan dust transport across Amazonia (and thus Ca supply) to Ecuadorian montane forests are linked to the ENSO cycle via the resulting changes of precipitation patterns in the region.

3.2 Amazonian biomass burning derived acid and nutrient deposition in the north Andean montane forest of Ecuador (Section C)

The temporal course of GOME NO_2 integrated column amounts paralleled those of the concentrations of total H$^+$, N, NO_3-N, Mn, and TOC in rainfall at my study site. This coincides with the observation that N, Mn, and TOC deposition rates were at the upper end of published values in similar ecosystems. Significant differences in element concentrations of rainfall between "fire" and "no fire" periods existed for H$^+$, N_{tot}, NO_3-N, DON, TOC, and Mn. Drier conditions during "fire" periods also resulted in significantly higher dry deposition of biomass burning-related elements. Consequently, total depositions of NO_3-N, N_{tot}, DON, H$^+$, and Mn were higher during "fire" than "no fire" periods Thus, biomass burning in Amazonia is a major driving factor of element deposition even to distant montane forests at the outer rim of Amazonia.

Almost all deposited H$^+$ was buffered in the canopy both during "fire" and "no fire" conditions. Total deposition of base metals tended to be lower during "fire" conditions than during "no fire" conditions, but base metal concentrations in throughfall were higher during "fire" than during "no fire" because of the enhanced H$^+$ buffering in the canopy during "fire". Elevated N deposition during "fire" periods increased concentrations in throughfall. Nevertheless, N was retained in the canopy during "fire" conditions but not during "no fire" conditions, when N deposition was much smaller (and therefore should even be more strongly retained if N was the only limiting nutrient). Two possible explanations for this observation exist: (i) forest growth at my study site is limited by several nutrients including N simultaneously ("co-limitation"), and (ii) N is not limiting but is taken up, if the limiting element is deposited at an increased rate because of the generally higher nutrient demand.

Elevated base metal leaching during "fire" conditions was also observed in the litter leachates below the organic horizon and to a lesser degree at 0.15 m depth in the mineral soil. The differences in total base metal loss from the 0.15 m depth in the mineral soil among transects (6.7-44 mol_c ha^{-1} wk^{-1}) can be attributed to the differences in base saturation among the transects (6.3-95%). Since 67% and 15% of the nutrient-absorbing roots are located in the O horizon and upper 0.15 m of the mineral soil, respectively [*Soethe et al.*, 2006], this loss is a nutrient depletion for the vegetation (in spite of retention in the subsoil).

At the 0.3 m depth and in stream water, there were no significant differences between "fire" and "no fire" conditions because of the large retention capacity of the mineral soil. Thus, the fire-derived acidification front reached, on average, the 0.15 m depth of the mineral soil between 1998 and 2003.

NO_3-N and N_{tot} fluxes in the O horizon also indicate N loss for all of the catchments. N losses were much higher during "no fire" conditions than during "fire" conditions, thus indicating (as in the canopy) that N is only retained during periods with generally elevated nutrient deposition.

Biomass burning intensity is linked to the ENSO [e. g., *Alencar et al.*, 2006; *van der Werf et al.*, 2006]. Taking the proposed frequency shift of future ENSO due to global warming into account [*Timmermann et al.*, 1999] as well as growing interest in soybean cropping [*Soares et al.*, 2006; *Arima et al.*, 2007], my findings suggests rising deposition rates of biomass burning-related aerosols to my study site in the future.

3.3 Water flow paths in soil control element exports in an Andean tropical montane forest (Section D)

Carbon, N, P, and S form two groups with respect to the relation between discharge and concentration. The first group, P and S, had maximum concentrations in stream water when discharge was low. In contrast, the second group, consisting of TOC and the N species, had maximum concentrations when discharge was high. The second group is subdivided into TOC, NH_4-N, DON with greater and NO_3-N with smaller concentrations during lateral flow, when the soil is waterlogged, than during stormflow. In particular, TOC and DON more than tripled their concentrations under "lateral flow" compared with "stormflow" conditions. The TOC, NO_3-N, NH_4-N, and DON concentrations in stream water are best explained by the assumption that rapid, near-surface flow carries elevated concentrations of these chemical constituents to the stream where the solutions mix with groundwater-fed baseflow. Besides the triggering of lateral flow during storm conditions, water saturation in soil reduces mineralization

[*Likens et al.*, 2002]. This would result in reduced N, P, and S release and reduced nitrification and thus explains the smaller S and NO_3-N concentrations under "lateral flow" compared to "storm flow" conditions. Furthermore, SO_4 and NO_3 are rapidly leached in the first discharge event after a drier period as a consequence of their mobility in soil.

The monitored metals can be classified into three groups in terms of their concentration-discharge level behaviour. The concentrations in stream water of the first group, K, Ca, and Mg, tended to increase continuously with increasing discharge levels. Aluminium and Mn formed the second group with almost constant concentrations during most discharge levels except for "lateral flow" where their concentrations in stream water increased by a factor of up to 20. In contrast to all other monitored metals, Na - the only representative of the third group of metals - showed a negative relation with discharge levels.

All elements had greater concentrations in the organic layer than in the mineral soil, but only C, N, K, Ca, Mg, Al and Mn were flushed out during lateral-flow conditions. Phosphorus, S and Na, in contrast, were mainly released by weathering and (re-)oxidation of sulfides in the subsoil. Baseflow accounted for 32% to 61% of P export, while >50% of S was exported during intermediate flow conditions, i.e. lateral flow at the depth of several tens of cm in the mineral soil.

For P and S, peak flow-related export was of little importance (<20% of the total export occurred during peak discharge conditions). In contrast to P and S, peak discharge accounted for approximately 50% of the export of TOC and the N species. For TOC, NH_4-N, and DON, the rarely occurring flow class "lateral flow" contributed as much to the export as did the more frequently occurring flow class "stormflow".

The extreme flow classes "storm" and "lateral" accounted for an estimated 50-75% of the total export of Al and Mn. These two flow classes also included around one third of the total export of K, Ca, Mg, and Na. My results demonstrate that storm-event related near-surface flow markedly affects the cycling of many nutrients in steep tropical montane forests. This might be of growing importance because Haylock *et al.* [2006] found a trend towards wetter and more extreme rainfall conditions for Ecuador by analyzing rainfall data from 1960 to 2000, and explained this by changes of ENSO.

3.4 Integrative statistical evaluation of the controls of Ca and N fluxes from rainfall through the forest ecosystem to surface runoff

Rainfall. The TOMS aerosol index above the study area is the single variable explaining the largest part of the variance of Ca concentrations in rainfall (Table A-2a). The site effect illustrated by the variables "MCs", which combine all properties of the three replicate microcatchments explains the second largest part followed by the sea surface temperature anomalies index "SST ENSO 3". This supports the findings described in Section B that Ca-rich aerosol transport from Saharan sources is related to ENSO. The highly significant interactions between "TOMS" and "SST ENSO 3" and between "SST ENSO 3" and "P transp. corridors" (Table A-2a) further emphasize the relation between aerosol transport and ENSO. Cold ENSO conditions ("La Niña"), generating negative values in the SST ENSO 3 index in combination with large aerosol loads at the study site highly significantly explain a considerable part of the variance of Ca concentrations in rainfall. The interaction between "SST ENSO 3" and rainfall occurring along the aerosol transport corridor above Amazonia ("P transp. corridors") illustrates that coinciding negative "SST ENSO 3" values ("La Niña") and high precipitation along the transport corridor allow the passage of Saharan aerosols over Amazonia because of the consequently higher frequency of dry spells during this time (see Section B). The variables "GOME" and "firepix" did not contribute significantly to explain the variance of Ca concentrations in rainfall showing that the contribution of biomass burning-related aerosol to the overall Ca deposition at the study site is negligible.

For Ca fluxes with rainfall, "TOMS" is again the single variable with the highest contribution to the explained variance followed by the site effect. Different to the concentrations, the variables "P h/wk" and P mm/h" representing rainfall intensity and frequency, respectively, have a small but significant effect. There were many significant interactions illustrating the complex interplay among Saharan dust transport, climatic conditions at my study site, and biomass burning in Amazonia (Table A-2b).

Rainfall intensity and frequency significantly explain the largest part of the variance of N_{tot} concentrations in rainfall followed by "SST ENSO 3" and "RF volume" (Table A-2c). This illustrates that mainly climatic variables influence N_{tot} concentrations in rainfall. The significant interaction between "long-range trajs" and "P h/wk" might indicate improved transport conditions for biomass-burning related N emissions in Amazonia to my study site where N is washed out from the atmosphere with rainfall.

Table A-2: Results of general linear models with a) Ca concentrations, b) Ca fluxes, c) N_{tot} concentrations, and d) N_{tot} fluxes in rainfall as target variables.

a)

	Source	df	SS	SS%	F	p
Single var.	MCs	2	0.34	6.6	7.4	***
	TOMS	1	0.43	8.2	18.4	***
	SST ENSO 3	1	0.09	1.8	3.9	*
	w)					
Interactions	TOMS * SST ENSO 3	1	0.70	13.5	30.2	***
	SST ENSO 3 * P transp. corridors	1	0.32	6.2	14.0	***
	error	112	2.60	50.1		

w) tested and not found significant: humidity, GOME, firepix, P h/wk, P mm/h, P transp. corridors, Long range trajs, windvel, RF volume

n.s. = not significant, *** p < 0.001, ** p < 0.01, * p < 0.05

b)

	Source	df	SS	SS%	F	p
Single var.	MCs	2	1126	5.0	7.9	***
	TOMS	1	1716	7.6	24.1	***
	P h/wk	1	1115	4.9	15.7	***
	P mm/h	1	297	1.3	4.2	*
	x)					
Interactions	TOMS * SST ENSO 3	1	1673	7.4	23.5	***
	TOMS * P h/wk	1	1097	4.8	15.4	***
	TOMS * P transp. corridors	1	960	4.2	13.5	***
	TOMS * Long range trajs	1	1359	6.0	19.1	***
	TOMS * firepix	1	1323	5.8	18.6	***
	TOMS * GOME	1	548	2.4	7.7	**
	Long range trajs * SST ENSO 3	1	1327	5.8	18.7	***
	Long range trajs * P mm/h	1	486	2.1	6.8	*
	error	109	7750	34.1		

x) tested and not found significant: humidity, windvel, P transp. corridors, Long range trajs, GOME, SST ENSO 3

n.s. = not significant, *** p < 0.001, ** p < 0.01, * p < 0.05

c)

	Source	df	SS	SS%	F	p
Single var.	MCs	2	0.03	0.3	0.3	n.s.
	P h/wk	1	2.37	22.4	49.3	***
	P mm/h	1	0.75	7.1	15.6	***
	SST ENSO 3	1	0.473	4.5	9.9	**
	y)					
Interactions	Long range trajs * P h/wk	1	0.24	2.3	5.1	*
	error	115	5.52	52.3		

y) tested and not found significant: TOMS, GOME, P transp. corridors, windvel, humidity

n.s. = not significant, *** p < 0.001, ** p < 0.01, * p < 0.05

d)

	Source	df	SS	SS%	F	p
Single var.	MCs	2	355	2.2	2.6	n.s.
	GOME	1	2874	17.7	41.4	***
	Long range trajs	1	1400	8.6	20.2	***
	z)					
Interactions	GOME * P mm/h	1	4905	30.2	70.7	***
	GOME * TOMS	1	1100	6.8	15.9	***
	GOME * P h/wk	1	931	5.7	13.4	***
	error	45	3122	19.2		

z) tested and not found significant: SST ENSO 3, humidity, P h/wk, TOMS, firepix

n.s. = not significant, *** p < 0.001, ** p < 0.01, * p < 0.05

Rainfall intensity and frequency significantly explain the largest part of the variance of N_{tot} concentrations in rainfall followed by "SST ENSO 3" and "RF volume" (Table A-2c). This illustrates that mainly climatic variables influence N_{tot} concentrations in rainfall. The significant interaction between "long-range trajs" and "P h/wk" might indicated improved transport conditions for biomass-burning related N emissions in Amazonia to my study site where N is washed out from the atmosphere with rainfall.

The single variable contributing most to the explained variance of the fluxes of N_{tot} in rainfall was "GOME" representing the NO_2 density seen as a column above the study site followed by long-range trajs (Table A-2d).

Throughfall. The site effect and "TF volume" together explain almost 60% of the variance of Ca concentrations in throughfall (Table A-3a). This illustrates that Ca concentrations in throughfall are strongly influenced by the properties of the soil the vegetation grows on, especially the base saturation. Furthermore, the dilution effect (i.e., the change in Ca concentration because of interception losses of rain water) controls Ca concentrations in throughfall. Further minor controls of Ca concentrations in throughfall include the Ca and H^+ concentrations in rainfall and climatic variables. The significant effect of Ca concentrations in rainfall on Ca concentrations in throughfall probably reflects the influence of Saharan dust-derived Ca inputs into my study ecosystem (see Section B) while the significant effect of H^+ concentrations in rainfall can be explained by the increased leaching of Ca because of the buffering of acids produced by biomass burning in Amazonia and transported to my study site (see Section C).

Similar to the Ca concentrations in throughfall, the Ca fluxes in throughfall are mainly controlled by the "site effect" and properties of rainfall. Interestingly, the Mn flux with rainfall has a small but significant effect supporting the view in Section C that Mn inputs might help overcome nutrient limitations resulting in enhanced Ca uptake in the canopy (Table A-3b). The H^+ fluxes with rainfall have only a significant impact at low rainfall conditions as reflected by the significant interaction between "P mm/h" and "RF flux H^+" possibly because only at low rainfall intensity a complete H^+ buffering resulting in increased Ca leaching occurs.

Table A-3: Results of general linear models with a) Ca concentrations, b) Ca fluxes, c) N_{tot} concentrations, and d) N_{tot} fluxes in throughfall as target variables.

a)

	Source	df	SS	SS%	F	p
Single var.	MCs	2	47.1	39.2	81.2	***
	TF volume	1	23.9	19.9	82.6	***
	RF conc H⁺	1	2.69	2.2	9.3	**
	humidity	1	2.29	1.9	7.9	**
	Temp ECSF	1	3.08	2.6	10.6	**
	SST ENSO 3	1	2.76	2.3	9.5	**.
	RF conc Ca	1	5.68	4.7	19.6	***
	TOMS	1	1.51	1.3	5.2	*
	radiation	1	3.52	2.9	12.2	***
	GOME	1	1.35	1.1	4.7	*
	w)					
Interactions	No interactions were significant					
	error	90	26.1	21.7		

w) tested and not found significant: RF volume, RF conc Mn, RF conc NO₃, RF conc N_{tot}, RF conc S, RF conc TOC, P h/wk, windvel, P mm/h

n.s. = not significant, *** p < 0.001, ** p < 0.01, * p < 0.05

b)

	Source	df	SS	SS%	F	p
Single var.	MCs	2	67899	68.4	169	***
	P mm/h	1	2770	2.8	13.8	***
	P h/wk	1	4643	4.7	23.1	***
	RF volume	1	2480	2.5	12.3	***
	RF flux Mn	1	1142	1.2	5.7	*
	TOMS	1	916	1.0	4.6	*.
	x)					
Interactions	P mm/h * RF flux H⁺	1	996	1.0	5.0	**
	error	63	12655	12.7		

x) tested and not found significant: SST ENSO 3, GOME, Temp ECSF, humidity, radiation, windvel, Tsoil0m Tsoil0.6m, soil water content 0.4m, RF flux H⁺, RF flux Ca, RF flux NO₃, RF flux N_{tot}, RF flux S, RF flux P, RF flux TOC

n.s. = not significant, *** p < 0.001, ** p < 0.01, * p < 0.05

c)

	Source	df	SS	SS%	F	p
Single var.	MCs	2	8.6	4.8	5.2	**
	TF volume	1	64.5	36.2	78.0	***
	Temp ECSF	1	18.5	10.4	22.4	***
	RF conc H⁺	1	11.0	6.1	13.2	***
	RF conc N_{tot}	1	3.37	1.9	4.1	*
	y)					
Interactions	No interactions were significant					
	error	78	64.8	36.2		

y) tested and not found significant: stormlateral, TOMS, SST ENSO 3, GOME, P h/wk, P mm/h, humidity, radiation, windvel, Tsoil0.6m, RF vol, RF conc Mn, RF conc S

n.s. = not significant, *** p < 0.001, ** p < 0.01, * p < 0.05

d)

	Source	df	SS	SS%	F	p
Single var.	MCs	2	10555	17.0	19.1	***
	RF flux N_{tot}	1	11784	19.0	42.6	***
	GOME	1	3017	4.9	10.9	**
	humidity	1	5757	9.3	20.8	***
	z)					
Interactions	RF flux N_{tot} * TOMS	1	2161	3.5	7.8	**
	error	62	17149	27.7		

z) tested and not found significant: TOMS, SST ENSO 3, P mm/h, P h/wk, radiation, windvel, Temp ECSF, RF flux Mn, Tsoil0m Tsoil0.6m, soil water content O₂, soil water content 0.4m, RF volume, RF flux H⁺, RF flux NO₃, RF flux S, RF flux P, RF flux TOC

n.s. = not significant, *** p < 0.001, ** p < 0.01, * p < 0.05

The "TF volume" explains more than half of total variance of N_{tot} concentrations in throughfall illustrating that N_{tot} concentrations are mainly driven by the dilution effect (Table A-3c). The significant effect of temperature on N_{tot} concentrations might be explained by enhanced plant growth (and subsequent N uptake). Alternatively, increased N deposition because of the increased fire activity in Amazonia and long-range transport to my study site in the warmer period of the year as described in Section C might be the process behind this effect. Further significant drivers include the H^+ and

N_{tot} concentrations in rainfall indicating that part of the deposited HNO_3 passes through the canopy and reaches the soil via throughfall. Finally, there is a significant but small site effect in spite of the small variation in soil N concentrations compared with the variation in soil Ca concentrations.

The N deposition with rainfall ("RF flux N_{tot}") is the single variable contributing most to the explained variance followed by the site effect, and air humidity (Table A-3d). This illustrates that there is little interaction of N in the canopy suggesting that my study system is not N-limited which would result in N retention. The NO_2 column above the study site ("GOME") significantly contributed to the explanation of the variance of N_{tot} fluxes with throughfall even if fitted after "RF flux N_{tot}" supporting the view that Amazonian forest fires are an important source of N deposition to my study site (see Section C). The only significant interaction between "RF flux N_{tot}" and "TOMS" is difficult to explain.

The following statistical analyses of soil solutions were only run for the concentrations because water fluxes were modeled with SWB and I wanted to avoid the potentially introduced bias of the statistical model because of the errors of the SWB model.

Litter leachate. Similar to throughfall, the site effect contributes most to the explained variance of Ca concentrations in litter leachate (Table A-4a). Thus, again the soil properties control Ca concentrations in litter leachate. The TOC concentrations in throughfall have the second strongest effect followed by the water content of the Oa horizon. Thus, organo-complexation and soil moisture play a key role in Ca export from the organic layer to the mineral soil as outlined in more detail in Section D. The significant effect of NO_3 concentrations in throughfall might point at a coupled export of Ca and NO_3 as postulated by *Perakis et al.* [2006]. The small but significant effects of "TOMS" and "SST ENSO 3" illustrate that the impact of Sahara-derived dust deposition still is visible in the Ca concentrations of the litter leachate. The significant effect of "TF conc. H^+" could again be explained by enhanced Ca leaching as a consequence of the Amazonian forest fire-derived acid deposition to my study site.

The N_{tot} concentration in throughfall contributes most to the explained variance of N_{tot} concentrations in litter leachate (Table A-4b). Thus most of the deposited N seems to be leached through the organic layer without much interaction further supporting my view that N is not limiting plant growth at my study site. The site effect is second largest and can be attributed to the differences in soil N concentrations and release rates. The third largest significant effect is that of "TOMS" which is unexpected and difficult to explain.

Table A-4: Results of general linear models with a) Ca and b) N_{tot} concentrations in litter leachate as target variable

a)

	Source	df	SS	SS%	F	p
Single var.	MCs	2	626	52.4	106.6	***
	TF conc TOC	1	211.7	17.7	72.0	***
	Soil water content O_e	1	59.3	5.0	20.2	***
	TF conc NO_3	1	33.2	2.8	11.3	**
	TOMS	1	13.7	1.1	4.7	*
	SST ENSO 3	1	16.9	1.4	6.0	*.
	TF conc H^+	1	18.3	1.5	6.5	*
	y)					
Interactions	No interactions were significant					
	error	76	213	17.8		

y) tested and not found significant: stormlateral, GOME, Temp ECSF, P h/wk, P mm/h, humidity, radiation, Tsoil 0m, RF volume, TF volume, LL volume, TF conc Ca, TF conc N_{tot}, TF conc S, TF conc Mn

n.s. = not significant, *** $p < 0.001$, ** $p < 0.01$, * $p < 0.05$

b)

	Source	df	SS	SS%	F	p
Single var.	MCs	2	194	16.6	17.5	***
	TF conc N_{tot}	1	474	40.5	85.7	***
	TOMS	1	33	2.8	6.0	*
	z)					
Interactions	No interactions were significant					
	error	77	426	36.4		

z) tested and not found significant: stormlateral, Gome, SST ENSO 3, T ECSF, P h/wk, P mm/h, humidity, radiation, windvel, Tsoil0m, RFvolume, TFvolume, TF conc Ca, TF conc H, TF conc TOC, Soil water contnet O_e, TF conc Mn, TF conc NO_3, LL volume

n.s. = not significant, *** $p < 0.001$, ** $p < 0.01$, * $p < 0.05$

Mineral soil solution. Similar to the litter leachate, most of the variance of Ca concentrations in mineral soil solution at 0.15 m soil depth is explained by the site effect and TOC concentrations in the input solution to the mineral soil (i.e., the litter leachate, Table A-5a). The third strongest effect has the soil temperature suggesting that biological turnover processes of organic matter play an important role for the Ca concentrations in the mineral soil solution at 0.15 m depth. The significant effect of NO_3^- concentrations in litter leachate on Ca concentrations in mineral soil solution at 0.15 m soil depth further supports the above stated assumption of a combined NO_3^- and Ca transport. Again similar to the litter leachate "TOMS" has a significant effect on Ca concentrations in mineral soil solution illustrating that the impact of Sahara-dust derived Ca impacts reaches deep into the soil. Finally, "TF volume" is a significant control of Ca concentrations in the mineral soil solution because of the dilution effect.

The NO_3^- concentrations in litter leachate contribute most to the explained variance of N_{tot} concentrations in mineral soil solution at 0.15m depth soil solution followed by the site effect and the temperature as driver of biological turnover processes releasing N (Table A-5b). The strong effect of TOC concentrations in litter leachate on N_{tot}

concentrations in mineral soil solution at 0.15 m depth illustrates the important contribution of organically bound N to total N concentrations in soil solution [*Goller et al.* 2006]. Furthermore, "TOMS" and "SST ENSO 3" had a significant effect on N_{tot} concentrations in mineral soil solution at 0.15 m depth again supporting the view of a linked Ca and NO_3^- transport. Finally, the Mn concentrations in litter leachate had a small but significant effect on N_{tot} concentrations in mineral soil solution at 0.15 m depth again suggesting that an increased availability of Mn might enhance N uptake by plants.

The GLMs to explain Ca and N_{tot} concentrations in mineral soil solution at 0.3 m depth resemble those at 0.15 m depth with strong site effects, dominating influences of the input concentrations (i.e. from 0.15 m soil depth) and a still visible influence of the Sahara-dust derived Ca inputs into the ecosystem (Table A-5c, d).

The large number of signficant interactions of the explanatory variables in all four GLMs used to explain Ca and N_{tot} concentrations in mineral soil solution at 0.15 and 0.3 m depths illustrate the complex processes occurring in soil.

Surface runoff. The site effect ("MCs") clearly contributes most to the runoff concentrations of Ca (Table A-6a). Thus, soil properties are the most important control of Ca export from my study catchments. Furthermore, "TOMS" and "SST ENSO 3" have significant effects illustrating that the increased Ca input during periods in which the transport of Sahara-derived dust is favored affects all ecosystem fluxes (see Section B). Finally, soil water contents influencing the water flow regime in soil have significant effects on Ca concentrations in runoff water illustrating the important role of flow paths in soil as outlined in Section D.

The strongest effect on N_{tot} concentrations in runoff has "GOME", i.e. the N concentration in the atmosphere above my study site followed by "RF conc. N_{tot}", TF volume as driver of soil water fluxes, and "RF conc. NO_3" (Table A-6). This supports the view that much of the introduced N passes through the whole ecosystem without being completely retained by the vegetation which would be expected in N-limited ecosystems. The significant effects of "TF conc. H^+", "SST ENSO 3" and "TF conc. Ca" indicate that the postulated coupled Ca and NO_3^- transport occurs through the whole ecosystem. Finally, "SS30 conc. Mn" has a significant effect furthermore supporting the view that increased Mn availability enhances N retention in the ecosystem.

Table A-5: Results of general linear models with a) Ca and b) N_{tot} concentrations at 0.15 m soil depth and c) Ca and d) N_{tot} concentrations at 0.3 m soil depths as target variables.

a)

	Source	df	SS	SS%	F	p
Single var.	MCs	2	146.5	24.3	35.5	***
	LL conc TOC	1	59.7	9.9	29.0	***
	Tsoil0m	1	49.7	8.2	24.1	***
	LL conc NO_3	1	31.9	5.3	15.5	***
	TOMS	1	29.2	4.8	14.2	***
	TF volume	1	29.4	4.9	14.3	***
	w)					
Interactions	Tsoil0m * LL conc NO_3	1	47.9	7.9	23.3	***
	TOMS * LL conc TOC	1	22.9	3.8	11.1	**
	P mm/h * LL conc TOC	1	19.8	3.3	9.6	**
	LL conc H^+ * LL conc TOC	1	20.3	3.4	9.9	**
	SST ENSO 3 * LL conc TOC	1	9.28	1.5	4.5	*
	error	46	94.8	15.7		

w) tested and not found significant: LL conc Mn, SST ENSO 3, GOME, LL conc N_{tot}, P h/wk, P mm/h, humidity, radiation, windvel, soil water content O_1, SS0.15m volume, LL conc Ca, LL conc H^+, Tsoil0.6m, LL volume, RF volume

n.s. = not significant, *** p < 0.001, ** p < 0.01, * p < 0.05

b)

	Source	df	SS	SS%	F	p
Single var.	MCs	2	75.5	11.8	22.5	***
	LL conc NO_3	1	88.7	13.8	52.9	***
	Tsoil0m	1	69.7	10.9	41.6	***
	TOMS	1	44.3	6.9	26.4	***
	RF volume	1	29.0	4.5	17.3	***
	LL conc TOC	1	54.7	8.5	32.6	***
	LL conc Mn	1	7.41	1.2	4.4	*
	x)					
Interactions	Tsoil0m * LL conc NO_3	1	102	15.9	60.8	***
	P mm/h * LL conc TOC	1	18.8	2.9	11.2	**
	SST ENSO 3 * Tsoil0m	1	17.5	2.7	10.4	**
	LL conc Mn * LL conc TOC	1	15.9	2.5	9.5	**
	error	59	98.9	15.4		

x) tested and not found significant: stormlateral, Gome, Temp Ecsf, P h/wk, humidity, radiation, windvel, Tsoil0.6m, soil water content O_1, soil water content 0.2m, TF volume, LL volume, LL conc H^+, LL conc Ca, LL conc N_{tot}

n.s. = not significant, *** p < 0.001, ** p < 0.01, * p < 0.05

c)

	Source	df	SS	SS%	F	p
Single var.	MCs	2	35.6	25.0	60.4	***
	SS15 conc Ca	1	48.0	33.4	163	***
	SS15 conc N_{tot}	1	7.92	5.6	26.9	***
	RF volume	1	7.69	5.4	26.1	***
	SS15 conc TOC	1	3.22	2.3	10.9	**
	y)					
Interactions	SS15 conc Ca * SS15 conc TOC	1	12.5	8.8	42.5	***
	TOMS * SS15 conc N_{tot}	1	3.22	2.3	10.9	**
	TOMS * SS15 conc Ca	1	2.34	1.6	8.0	**
	TOMS * Tsoil0.6m	1	1.46	1.0	5.0	*
	error	60	17.7	12.4		

y) tested and not found significant: stormlateral, GOME, SST ENSO 3, Temp ECSF, P h/wk, P mm/h, Tsoil0m, soil water content O_1, TF volume, soil water content 0.2m, SS30 volume, SS15 conc H^+, SS15 conc Mn, SS 5 conc NO_3

n.s. = not significant, *** p < 0.001, ** p < 0.01, * p < 0.05

d)

	Source	df	SS	SS%	F	p
Single var.	MCs	2	8.40	9.0	7.8	**
	SS15 conc NO_3	1	24.4	26.3	45.2	***
	TOMS	1	6.15	6.6	11.4	**
	SS15 conc Mn	1	3.49	3.8	6.5	*
	Tsoil0.6m	1	2.76	3.0	5.1	*
	z)					
Interactions	soil water content 0.2m * SS15 conc NO_3	1	3.42	3.7	6.3	*
	RF volume * SS15 conc Mn	1	2.77	3.0	5.1	*
	error	59	31.9	34.3		

z) tested and not found significant: stormlateral, GOME, SST ENSO 3, P mm/h, Temp ECSF, Tsoil0m, soil water content O_1, TF volume, SS30 volume, SS15 conc H^+, SS15 conc Ca, SS15 conc TOC, SS15 conc N_{tot}

n.s. = not significant, *** p < 0.001, ** p < 0.01, * p < 0.05

Table A-6: Results of general linear models with a) C_a and b) N_{tot} concentrations in runoff as target variables.

a)

	Source	df	SS	SS%	F	p
Single var.	MCs	2	1.17	32.3	26.8	***
	TOMS	1	0.26	7.1	11.7	**
	SST ENSO 3	1	0.18	5.0	8.3	**
	Soil water content 0.2m	1	0.17	4.8	8.0	**
	Soil water content O_z	1	0.28	7.7	12.7	***
	y)					
Interactions	No interactions were significant					
	error	63	1.38	38.1		

y) tested and not found significant: stormlateral, GOME, P transp. corridors, P h/wk, P mm/h, Tsoil0m Tsoil0.6m, Temp ECSF, soil water content 0.4m, RF volume, RF conc H⁺, RF conc Ca, RF conc Mn, RF conc NO_3, RF conc N_{tot}, RF conc TOC, TF volume, TF conc H⁺, TF conc Ca, TF conc NO_3, TF conc N_{tot}, TF conc TOC, TF conc H⁺, SS30 conc Ca, SS30 conc NO_3, SS30 conc NO_3, SS30 conc N_{tot}, SS30 conc TOC, SF volume

n.s. = not significant, *** p < 0.001, ** p < 0.01, * p < 0.05

b)

	Source	df	SS	SS%	F	p
Single var.	MCs	2	0.02	0.5	0.7	n.s.
	GOME	1	0.69	16.2	45.4	***
	RF conc N_{tot}	1	0.44	10.4	29.1	***
	TF volume	1	0.63	14.8	41.6	***
	RF conc NO_3	1	0.49	11.7	32.7	***
	TF conc H⁺	1	0.21	5.0	14.1	***
	SST ENSO 3	1	0.14	3.2	9.0	**
	TF conc Ca	1	0.14	3.2	9.0	**
	SS30 conc Mn	1	0.07	1.6	4.5	*
	z)					
Interactions	No interactions were significant					
	error	57	0.86	20.4		

z) tested and not found significant: TOMS, P transport corridors, TempECSF, P h/wk, P mm/h, Tsoil0m, Tsoil0.6m, stormlateral, soil water content O_z, soil water content 0.2m, soil water content 0.4m, TF conc NO_3, TF conc N_{tot}, TF conc TOC, SS30 conc H⁺, SS30 conc Ca, SS30 conc NO_3, SS30 conc Ntot, SS30 conc TOC, SF volume, RF conc H⁺, RF conc Ca, RF conc TOC

n.s. = not significant, *** p < 0.001, ** p < 0.01, * p < 0.05

3.5 Error discussion

The presented results and derived interpretations include some uncertainties. Dealing with climate-related phenomena like long-range transport of aerosols, trade wind systems or even climate variations like ENSO or SST almost instantly leads to large uncertainties, as interpretations and assumptions never can be supported by the necessary data base, which does not exist. Whenever I assessed climate data on my own, for example in case of identifying precipitation above Amazonia, I did this with the necessary precaution, well knowing that 24 meteorological stations made available by INPE can not display the whole reality. Lagrangian trajectory analyses usually have a large uncertainty, which I tried to compensate by calculating trajectories both backward and forward and only using them if the derived information was consistent. Another uncertainty is associated with the used satellite data, since detection of aerosols, fire pixels or NO_2 from space is challenged in various ways by e.g., cloud cover, to name the most common. In all cases, I used only data verified by the remote sensing specialists running the sensors themselves. Furthermore all conclusions based on these data are of comparative nature and therefore only represent trends.

Another uncertainty related to the special obstacles in climate research is the long-term character of the related phenomena. I am therefore fully aware of the problems arising from drawing conclusions on climate oscillations like e.g., ENSO having only data of five years, but when done these conclusions were always marked as hypotheses.

For the chemical part of my study addressing the uncertainties is easier. Concentrations of analyzed elements are partly low. Detection limits are given in Section C, p. 68. I treated values below detection limit as zero leading to underestimation of the real mean concentrations or fluxes. The sampling interval of the ecosystem solutions was one week. I did not poison the sample collectors in the field. During the time between the sampling intervals, the chemical composition of the solutions might change due to biological and chemical transformations. For instance, microbes could catalyze the reaction of organic N to inorganic N. However, also losses could occur, e.g. if denitrification led to gaseous N species like NO_x, N_2O, or N_2. This was excluded by comparing the weekly sampled solution to daily sampled solution. There were no significant differences between the calculated mean weekly chemical composition of the daily collected samples and the measured chemical composition of the cumulative weekly sample. Precipitation collectors have an open top, thus evaporation could occur. I used the table tennis ball to lower evaporation.

Total N and P in solution are determined by oxidation with $K_2S_2O_8$ and UV radiation. This method may underestimate total N and P concentrations particularly if recalcitrant N- or P-containing organic compounds that are not easily oxidizable are present in the solution [*Kaiser et al.*, 2003]. I was not able to control this potential error but expect that it affected all samples to a similar extent so that my comparisons among the analyzed samples are not substantially biased.

Dry deposition was estimated using the model of Ulrich [1983]. In this model, the ratio between throughfall and bulk deposition of Na or Cl⁻ (deposition ratio) is used for calculating dry deposition. Negligible interactions between the canopy and Na and Cl⁻ ions are assumed. Therefore, these elements were considered as inert tracers for the input of other elements or compounds. For two reasons these assumptions can be criticized: (1) the deposition of other elements and compounds might depend on other factors than those controlling Na and Cl⁻ deposition, and (2) Na and Cl⁻ ions might be taken up by or leached from the canopy. Indeed, I observed a sink for Na in the canopy, possibly attributable to phyllospheric cyanophytes, which are abundant on the tree leaves at my study site, and are known to require Na as an essential nutrient [*Freiberg*, 1999]. Therefore, I used Cl⁻ for which there were no indications of significant canopy interactions as a tracer.

4. Conclusions

1) Hypotheses i-iii (Section B): The Amazon basin does not necessarily act as the final sink for Saharan dust in spite of its high precipitation. Tropical montane rain forest in south Ecuador receives substantial contributions of base metals from north African sources. This base metal deposition was large enough to change the overall Ca and Mg budgets in the north Andean montane forests. Increased base metal deposition was related to dust outbursts of the Sahara and an Amazonian precipitation pattern with trans-regional dry spells allowing for dust transport to the Andes. The increased base metal deposition coincided with a strong La Niña event in 1999/2000.

2) Hypotheses iv, v (Section C): I observed a substantially elevated deposition of H^+, NO_3, N_{tot}, DON, Mn, and TOC during the biomass burning season in Amazonia and upwind which was at the upper end of published values for similar ecosystems except for H^+. Elevated H deposition during biomass burning caused elevated base metal loss from the canopy and the organic horizon and deteriorated already low base metal supply of the vegetation. N was only retained during biomass burning but not during non-fire conditions when deposition was much smaller. I conclude that biomass burning-related aerosol emissions in Amazonia are large enough to substantially increase element deposition at the western rim of Amazonia. Particularly the related increase of acid deposition impoverishes already base-metal scarce ecosystems. As biomass burning is most intense during El Niño situations, a shortened ENSO cycle because of global warming likely enhances the acid deposition at my study forest.

3) Hypotheses vi-iv (Section D): Peak C, N, K, Ca, Mg, Al, and Mn concentrations in stream water were associated with lateral flow (fast near-surface flow in saturated topsoil) while the greatest P, S, and Na concentrations occurred during low baseflow conditions. All elements had greater concentrations in the organic layer than in the mineral soil, but only C, N, K, Ca, Mg, Al and Mn were flushed out during lateral-flow conditions. Phosphorus, S and Na, in contrast, were mainly released by weathering and (re-) oxidation of sulfides in the subsoil. Baseflow accounted for 32% to 61% of P export, while >50% of S was exported during intermediate flow conditions, i.e. lateral flow at the depth of several tens of cm in the mineral soil. Near-surface water flow through C- and nutrient-rich topsoil during rainstorms was

the major export pathway for C, N, Al, and Mn (contributing >50% to the total export of these elements). Near-surface flow also accounted for one third of total base metal export. These results demonstrate that storm-event related near-surface flow markedly affects the cycling of many nutrients in steep tropical montane forests.

5. References

Alencar, A., D. Nepstad, and M. D. V. Diaz (2006), Forest understory fire in the Brazilian Amazon in ENSO and non-ENSO years: Area burned and committed carbon emissions, *Earth Interactions, 10*, doi:10.1175/EI1150.1171.

Allen, A. G., and A. H. Miguel (1995), Biomass Burning in the Amazon - Characterization of the Ionic Component of Aerosols Generated Tom Flaming and Smoldering Rain-Forest and Savanna, *Environmental Science & Technology, 29*(2), 486-493.

Andreae, M. O., et al. (1988), Biomass-burning emissions and associated haze layers over Amazonia, *Journal of Geophysical Research 93*(D2), 1509-1527.

Artaxo, P., J. V. Martins, M. A. Yamasoe, A. S. Procopio, T. M. Pauliquevis, M. O. Andreae, P. Guyon, L. V. Gatti, and A. M. C. Leal (2002), Physical and chemical properties of aerosols in the wet and dry seasons in Rondonia, Amazonia, *Journal of Geophysical Research-Atmospheres, 107*(D20), -.

Balslev, H., and B. Ollgaard (2002), Mapa de vegetación del sur de Ecuador, in *Botánica Austroecuatoriana. Estudios sobre los recursos vegetales en las provincias de El Oro, Loja y Zamora-Chinchipe*, edited by M. Z. Aguirre, et al., pp. 51-64, Ediciones Abya-Yala, Quito, Ecuador.

Bendix, J., R. Rollenbeck, D. Gottlicher, and J. Cermak (2006), Cloud occurrence and cloud properties in Ecuador, *Climate Research, 30*(2), 133-147.

Beven, K. J., R. Lamb, P. F. Quinn, and R. B. Romanowicz (1995), TOPMODEL, in *Computer Models of Watershed Hydrology*, edited by V. P. Singh, pp. 627-668, Water Resource Publications.

Brooks, T. M., et al. (2002), Habitat loss and extinction in the hotspots of biodiversity, *Conservation Biology, 16*(4), 909-923.

Bruijnzeel, L. A., and L. S. Hamilton (2000), Decision time for cloud forests, IHP-Unesco and WWF International, Paris, Amsterdam.

Buffam, I., J. N. Galloway, L. K. Blum, and K. J. McGlathery (2001), A stormflow/baseflow comparison of dissolved organic matter concentrations and bioavailability in an Appalachian stream, *Biogeochemistry, 53*(3), 269-306.

Campbell, J. L., J. W. Hornbeck, W. H. McDowell, D. C. Buso, J. B. Shanley, and G. E. Likens (2000), Dissolved organic nitrogen budgets for upland, forested ecosystems in New England, *Biogeochemistry, 49*(2), 123-142.

Claquin, T., M. Schulz, and Y. J. Balkanski (1999), Modeling the mineralogy of atmospheric dust sources, *Journal of Geophysical Research-Atmospheres, 104*(D18), 22243-22256.

Clark, K. L., N. M. Nadkarni, D. Schaefer, and H. L. Gholz (1998), Atmospheric deposition and net retention of ions by the canopy in a tropical montane forest, Monteverde, Costa Rica, *Journal of Tropical Ecology, 14*, 27-45.

Cuevas, E., and E. Medina (1986), Nutrient dynamics within Amazonian forests I. Nutrient flux in fine litter fall and efficiency of nutrient utilization, *Oecologia*, *68*, 466-472.

Cuevas, E., and E. Medina (1988), Nutrient dynamics within amazonian forests II. Fine root growth, nutrient availability and litter decomposition, *Oecologia*, *76*, 222-235.

Da Rocha, G. O., A. G. Allen, and A. A. Cardoso (2005), Influence of agricultural biomass burning on aerosol size distribution and dry deposition in southeastern Brazil, *Environmental Science & Technology*, *39*(14), 5293-5301.

Draxler, R. R., and G. D. Hess (1998), An overview of the HYSPLIT_4 modelling system for trajectories, dispersion and deposition, *Australian Meteorological Magazine*, *47*(4), 295-308.

DVWK (Deutscher Verband für Wasserwirtschaft und Kulturbau) (1996), *Ermittlung der Verdunstung von Land- und Wasserflächen*, DVWK, Bonn.

Evans, R. D., I. F. Jefferson, R. Kumar, K. O'Hara-Dhand, and I. J. Smalley (2004), The nature and early history of airborne dust from North Africa; in particular the Lake Chad basin, *Journal of African Earth Sciences*, *39*(1-2), 81-87.

Fleischbein, K., W. Wilcke, R. Goller, J. Boy, C. Valarezo, W. Zech, and K. Knoblich (2005), Rainfall interception in a lower montane forest in Ecuador: effects of canopy properties, *Hydrological Processes*, *19*(7), 1355-1371.

Fleischbein, K., W. Wilcke, C. Valarezo, W. Zech, and K. Knoblich (2006), Water budgets of three small catchments under montane forest in Ecuador: experimental and modelling approach, *Hydrological Processes*, *20*(12), 2491-2507.

Freiberg, E. (1999), Influence of microclimate on the occurrence of cyanobacteria in the phyllosphere in a premontane rain forest of Costa Rica, *Plant Biology*, *1*(2), 244-252.

Godsey, S., H. Elsenbeer, and R. Stallard (2004), Overland flow generation in two lithologically distinct rainforest catchments, *Journal of Hydrology*, *295*(1-4), 276-290.

Goller, R., W. Wilcke, M. J. Leng, H. J. Tobschall, K. Wagner, C. Valarezo, and W. Zech (2005), Tracing water paths through small catchments under a tropical montane rain forest in south Ecuador by an oxygen isotope approach, *Journal of Hydrology*, *308*(1-4), 67-80.

Goller, R., W. Wilcke, K. Fleischbein, C. Valarezo, and W. Zech (2006), Dissolved nitrogen, phosphorus, and sulfur forms in the ecosystem fluxes of a montane forest in ecuador, *Biogeochemistry*, *77*(1), 57-89.

Gustafsson, J. P., D. Berggren, M. Simonsson, M. Zysset, and J. Mulder (2001), Aluminium solubility mechanisms in moderately acid Bs horizons of podzolized soils, *European Journal of Soil Science*, *52*(4), 655-665.

Haylock, M. R., et al. (2006), Trends in total and extreme South American rainfall in 1960-2000 and links with sea surface temperature, *Journal of Climate*, *19*(8), 1490-1512.

Hedin, L. O., P. M. Vitousek, and P. A. Matson (2003), Nutrient losses over four million years of tropical forest development, *Ecology*, *84*(9), 2231-2255.

Homeier, J. (2004), Baumdiversität, Waldstruktur und Wachstumsdynamik zweier tropischer Bergregenwälder in Ecuador und Costa Rica, Ph. D thesis, University of Bielefeld, Bielefeld, Germany.

Hook, A. M., and J. A. Yeakley (2005), Stormflow dynamics of dissolved organic carbon and total dissolved nitrogen in a small urban watershed, *Biogeochemistry*, *75*(3), 409-431.

Hungerbühler, D. (1997), Neogene basins in the Andes of southern Ecuador: evolution, deformation and regional tectonic implications, Ph. D. thesis, ETH, Zürich.

Kaiser, K., G. Guggenberger, and L. Haumaier (2003), Organic phosphorus in soil water under a European beech (Fagus sylvatica L.) stand in northeastern Bavaria, Germany: seasonal variability and changes with soil depth, *Biogeochemistry*, *66*(3), 287-310.

Kauffman, J. B., D. L. Cummings, D. E. Ward, and R. Babbitt (1995), Fire in the Brazilian Amazon .1. Biomass, Nutrient Pools, and Losses in Slashed Primary Forests, *Oecologia*, *104*(4), 397-408.

Kaufman, Y. J., I. Koren, L. A. Remer, D. Tanre, P. Ginoux, and S. Fan (2005), Dust transport and deposition observed from the Terra-Moderate Resolution Imaging Spectroradiometer (MODIS) spacecraft over the Atlantic ocean, *Journal of Geophysical Research-Atmospheres*, *110*(D10), -.

Likens, G. E., and J. S. Eaton (1970), A polyurethane stemflow collector for trees and shrubs, *Ecology*, *51*, 938-939.

Likens, G. E., C. T. Driscoll, D. C. Buso, M. J. Mitchell, G. M. Lovett, S. W. Bailey, T. G. Siccama, W. A. Reiners, and C. Alewell (2002), The biogeochemistry of sulfur at Hubbard Brook, *Biogeochemistry*, *60*(3), 235-316.

Lloyd, C. R., and A. Marques (1988), Spatial variability of throughfall and stemflow measurements in Amazonian rain forest, *Agricultural and Forest Meteorology*, *47*, 63-73.

Mahowald, N. M., P. Artaxo, A. R. Baker, T. D. Jickells, G. S. Okin, J. T. Randerson, and A. R. Townsend (2005), Impacts of biomass burning emissions and land use change on Amazonian atmospheric phosphorus cycling and deposition, *Global Biogeochemical Cycles*, *19*(4), -.

Matzner, E. (2004), *Biogeochemistry of Forested Catchments in a changing Environment. A German case study*, Springer, Berlin.

McDowell, W. H., and C. E. Asbury (1994), Export of Carbon, Nitrogen, and Major Ions from 3 Tropical Montane Watersheds, *Limnology and Oceanography*, *39*(1), 111-125.

McLaughlin, S. B., and R. Wimmer (1999), Tansley Review No. 104 - Calcium physiology and terrestrial ecosystem processes, *New Phytologist*, *142*(3), 373-417.

McPhaden, M. J. (1999), Genesis and evolution of the 1997-98 El Nino, *Science*, *283*(5404), 950-954.

Moreno, T., X. Querol, S. Castillo, A. Alastuey, E. Cuevas, L. Herrmann, M. Mounkaila, J. Elvira, and W. Gibbons (2006), Geochemical variations in aeolian mineral particles from the Sahara-Sahel Dust Corridor, *Chemosphere*, *65*(2), 261-270.

Okin, G. S., N. Mahowald, O. A. Chadwick, and P. Artaxo (2004), Impact of desert dust on the biogeochemistry of phosphorus in terrestrial ecosystems, *Global Biogeochemical Cycles*, *18*(2), -.

Parker, G. G. (1983), Throughfall and stemflow in the forest nutrition cycle, *Advances in Ecological Research*, *13*, 57-133.

Perakis, S., D. Maguire, T. Bullen, K. Cromack, R. Waring, and J. Boyle (2006), Coupled nitrogen and calcium cycles in forests of the Oregon Coast Range, *Ecosystems*, *9*(1), 63-74.

Pereira, E. B., A. W. Setzer, F. Gerab, P. E. Artaxo, M. C. Pereira, and G. Monroe (1996), Airborne measurements of aerosols from burning biomass in Brazil related to the TRACE a experiment, *Journal of Geophysical Research-Atmospheres*, *101*(D19), 23983-23992.

Prospero, J. M., and P. J. Lamb (2003), African droughts and dust transport to the Caribbean: Climate change implications, *Science*, *302*(5647), 1024-1027.

Reichholf, J. H. (1986), Is saharan dust a major source of nutrients for the amazonian rainforest *Studies on Neotropical Fauna and Environment 21*, 251-255.

Reid, E. A., J. S. Reid, M. M. Meier, M. R. Dunlap, S. S. Cliff, A. Broumas, K. Perry, and H. Maring (2003), Characterization of African dust transported to Puerto Rico by individual particle and size segregated bulk analysis, *Journal of Geophysical Research-Atmospheres*, *108*(D19), -.

Richter, A., J. P. Burrows, H. Nuss, C. Granier, and U. Niemeier (2005), Increase in tropospheric nitrogen dioxide over China observed from space, *Nature*, *437*(7055), 129-132.

Richter, M. (2003), Using epiphytes and soil temperatures for eco-climatic interpretations in Southern Ecuador, *Erdkunde*, *57*, 161-181.

Rieuwerts, J. S. (2007), The mobility and bioavailability of trace metals in tropical soils: a review, *Chemical Speciation and Bioavailability*, *19*(2), 75-85.

Rollenbeck, R., P. Fabian, and J. Bendix (2008), Temporal Heterogeneities - Matter deposition from remote areas, in *Gradients in a tropical mountain ecosystem of Ecuador*, edited, pp. 303-310.

Saunders, T. J., M. E. McClain, and C. A. Llerena (2006), The biogeochemistry of dissolved nitrogen, phosphorus, and organic carbon along terrestrial-aquatic flowpaths of a montane headwater catchment in the Peruvian Amazon, *Hydrological Processes*, *20*(12), 2549-2562.

Schellekens, J., F. N. Scatena, L. A. Bruijnzeel, A. I. J. M. van Dijk, M. M. A. Groen, and R. J. P. van Hogezand (2004), Stormflow generation in a small rainforest catchment in the luquillo experimental forest, Puerto Rico, *Hydrological Processes*, *18*(3), 505-530.

Schütz, L., R. Jaenicke, and H. Pietrek (1981), Saharan dust transport over the North Atlantic Ocean, *Geological Society of America, special paper 186*, 87-100.

Soethe, N., J. Lehmann, and C. Engels (2006), The vertical pattern of rooting and nutrient uptake at different altitudes of a south ecuadorian montane forest, *Plant and Soil*, *286*(1-2), 287-299.

Swap, R., S. Garstang, S. Greco, R. Talbot, and P. Kallberg (1992), Saharan dust in the Amazon basin, *Tellus*, *44*(B), 133-149.

Timmermann, A., J. Oberhuber, A. Bacher, M. Esch, M. Latif, and E. Roeckner (1999), Increased El Nino frequency in a climate model forced by future greenhouse warming, *Nature*, *398*(6729), 694-697.

Torres, O., P. K. Bhartia, J. R. Herman, A. Sinyuk, P. Ginoux, and B. Holben (2002), A long-term record of aerosol optical depth from TOMS observations and comparison to AERONET measurements, *Journal of the Atmospheric Sciences*, *59*(3), 398-413.

Ulrich, B. (1983), *Effects of Accumulation of Air Pollutants in Forest Ecosystems*, Reidel, Dordrecht.

USDA-NRCS, U. S. D. o. A.-N. R. C. S. (1998), *Keys to Soil Taxonomy*, Pocahontas Press, Washington DC.

van der Werf, G. R., J. T. Randerson, L. Giglio, G. J. Collatz, P. S. Kasibhatla, and A. F. Arellano (2006), Interannual variability in global biomass burning emissions from 1997 to 2004, *Atmospheric Chemistry and Physics*, *6*, 3423-3441.

Whitehead, H. L., and J. H. Feth (1964), Chemical composition of rain, dry fallout and bulk precipitation at Menlo Park, California, *Journal of Geophysical Research*, *69*, 3319-3333.

Wilcke, W., S. Yasin, C. Valarezo, and W. Zech (2001), Change in water quality during the passage through a tropical montane rain forest in Ecuador, *Biogeochemistry*, 55(1), 45-72.

Wilcke, W., S. Yasin, U. Abramowski, C. Valarezo, and W. Zech (2002), Nutrient storage and turnover in organic layers under tropical montane rain forest in Ecuador, *European Journal of Soil Science*, 53(1), 15-27.

Wilcke, W., H. Valladarez, R. Stoyan, S. Yasin, C. Valarezo, and W. Zech (2003), Soil properties on a chronosequence of landslides in montane rain forest, Ecuador, *Catena*, 53(1), 79-95.

B Tropical Andean forest derives calcium and magnesium from Saharan dust[a]

Jens Boy[1,2] & Wolfgang Wilcke[2]

[1] Institute of Ecology, Berlin University of Technology, Salzufer 11-12, 10587 Berlin, Germany

[2] Geographic Institute, Johannes Gutenberg University, 55099 Mainz, Germany

[a] Global Biogeochemical Cycles, Vol. 22, GB1027, doi:10.1029/2007GB002960, 2008

1. Abstract

We quantified base metal deposition to Amazonian montane rain forest in Ecuador between May 1998 and April 2003 and assessed the response of the base metal budget of three forested microcatchments (8-13 ha). There was a strong interannual variation in deposition of Ca [4.4-29 kg ha^{-1} a^{-1}], Mg [1.6-12], and K [9.8-30]). High deposition changed the Ca and Mg budgets of the catchments from loss to retention, suggesting that the additionally available Ca and Mg was used by the ecosystem. Increased base metal deposition was related to dust outbursts of the Sahara and an Amazonian precipitation pattern with trans-regional dry spells allowing for dust transport to the Andes. The increased base metal deposition coincided with a strong La Niña event in 1999/2000.

Abbreviations: ENSO, El Niño Southern Oscillation; MC, microcatchment; CFA, continous flow analyzer; TDN, total dissolved nitrogen; TDS, total dissolved sulphur; TD, total deposition; TFD, throughfall deposition; RD, rainfall deposition; DD, dry deposition; SFD, stemflow deposition; LEA, canopy budget; VWM, volume weighted mean; FWM, flow weighted mean; TOMS, 'Total Ozone Mapping Spectrometer'; HYSPLIT, 'Hybrid Single Particle Lagrangian Integrated Trajectories'; ARL, 'Air Resource Laborartory'; INPE, 'Instituto Nacional de Pesquisas Espaciais'; AI, aerosol index

2. Introducion

Tropical montane rain forests still belong to the least explored ecosystems of the world [*Hamilton et al.*, 1995]. At the same time these ecosystems are among the plant and animal richest of the world and now threatened to vanish [*Brooks et al.*, 2002]. To develop long-term conservational strategies understanding of the ecosystem functioning is crucial. This requires investigating the controls of forest nutrition and their response to long-term variations in environmental constraints, e.g. atmospheric inputs. In ecosystems where nutrient demand of the vegetation is not covered by weathering of the parent rock, atmospheric inputs are an important source, and might help to overcome nutrient limitations [*Hedin et al.*, 2003].

The composition of atmospheric input is a function of source region. Besides sea spray, local dust, volcanism, and biomass burning the most important source of atmospheric input to Amazonian forests is the Sahara desert [*Kaufman et al.*, 2005; *Reichholf*, 1986; *Swap et al.*, 1992]. The Sahara produces 400-700 Tg of atmospheric dust per year, which represents almost 50% of the global dust production [*Prospero and Lamb*, 2003; *Schütz et al.*, 1981]. A large portion of this dust is transported across the North Atlantic by the predominant westerly winds (240 ± 80 Tg [*Kaufman et al.*, 2005]). Although part (50 Tg) is deposited in the Caribbean, a significant portion (50 Tg) reaches the Amazon basin according to remote sensing [*Chiapello et al.*, 2005; *Kaufman et al.*, 2005]. By direct ground-based measurements, Saharan dust has been recorded as far inland the Amazon basin as Manaus [*Formenti et al.*, 2001; *Swap et al.*, 1992].

It has been hypothesized but never directly shown that desert dust helps reducing nutrient limitations in the Amazonian rain forests [*Okin et al.*, 2004; *Reichholf*, 1986; *Swap et al.*, 1992]. Among the elements considered to be potentially fertilizing are base metals (i.e., K, Ca and Mg), which can be abundant in Saharan dust [*Claquin et al.*, 1999; *Moreno et al.*, 2006; *Reid et al.*, 2003]. Base metal supply is particularly low in Amazonian ecosystems because of highly weathered soils and the absence of local base metal sources [*Cuevas and Medina*, 1986; *Swap et al.*, 1992; *Wilcke et al.*, 2001].

Base metals are important plant nutrients controlling key plant functions. Calcium plays an important role for growth and functioning of woody tissue, cell synthesis, physiological signaling or N uptake[*Likens et al*, 1998; *McLaughlin and Wimmer*, 1999; *Perakis et al.*, 2006]. Magnesium is an essential constituent of many cellular enzymes and the chlorophyll molecule. Sufficient supply of Mg is a prerequisite for aggregation of ribosomes and regulation of the ion balance in the cell[*Campo et al.*, 2000; *Shaul*, 2002]. Potassium is required as a major electrolyte and modulates osmotic balance in plant tissues. K plays also an important role in photosynthesis and the formation of roots [*Likens et al.*, 1994; *Tripler et al.*, 2006].

It is, however, unclear if Saharan dust can be transported in significant quantities across the usually humid Amazon basin towards the remoter parts of Amazonia and the montane forests of the Andes. As aerosols are almost instantly scavenged by rain, the Amazon basin should act as a wet barrier for Sahara dust transport. Therefore, Saharan dust can only reach remote Amazonian forests at the outer rim of the Amazon basin during dry spells allowing for dust transport across the region. The precipitation of the Amazon basin is linked to the El Niño Southern Oscillation (ENSO), being dryer during El Niño and wetter during La Niña conditions [*Zeng*, 1999]. This suggests that Saharan dust transport across Amazonia is more likely during warm ENSO phases. In addition, aerosol emission activity of the Sahara is at its maximum during warm ENSO phases [*Prospero and Lamb*, 2003].

To explore the role of atmospheric input for forest functioning of remote Amazonian forests we used a whole-catchment approach [*Bruijnzeel*, 1991] in an Andean lower tropical montane rain forest to calculate net element budgets at the ecosystem scale (including vegetation, organic layer, and mineral soil). We monitored all ecosystem fluxes for five consecutive years between 1998 and 2003. Our hypotheses were (i) Saharan dust transport adds significant amounts of base metals to montane forests at the western rim of the Amazon basin, (ii) the nutrient deposition via Saharan dust influences the catchment budget of base metals in an Ecuadorian montane forest, and (iii) dust transport across Amazonia is related to the precipitation pattern controlled by ENSO and therefore is an overseasonal phenomenon.

3. Material and Methods

3.1 Study site

The study area is located on the eastern slope of the "Cordillera Real", the eastern Andean cordillera in south Ecuador facing the Amazon basin at 4° 00` S and 79° 05` W. We selected three 30-50° steep and 8-13 ha large microcatchments (MC1-3) under montane forest at an altitude of 1900-2200 m above sea level (a.s.l.) for our study

(Figure A-1). We installed our equipment in each MC on transects, about 20 m long with an altitude range of 10 m, on the lower part of the slope at 1900-1910 m a.s.l. (transects MC1, MC2.1, and MC3). Moreover, we installed extra instrumentation at 1950-1960 (MC2.2) and 2000-2010 m a.s.l. (MC2.3). All transects were located below closed forest canopy and aligned downhill. Three unforested sites near these microcatchments were used for rainfall gauging. Gauging site 2 existed since April 1998, gauging sites 1 and 3 were built in May 2000. All catchments drain via small tributaries into the Rio San Francisco which flows into the Amazon basin.

Within the monitored period between April 1998 and April 2003 annual precipitation ranged between 2340 and 2667 mm. Additional climate data were available from a meteorological station [*Richter*, 2003] between MC 2 and 3 (Figure A-1). June tended to be the wettest month with 302 mm of precipitation on average, in contrast to 78 mm in each of November and January, the driest months. The mean temperature at 1950 m a.s.l. was 15.5 °C. The coldest month was July, with a mean temperature of 14.5 °C, the warmest November with a mean temperature of 16.6 °C.

Recent soils have developed on postglacial landslides or possibly from periglacial cover beds [*Wilcke et al.*, 2001; 2003]. Soils are Humic Eutrudepts on transect MC1, Humic Dystrudepts on transects MC2.1, MC2.2, and MC2.3, and Oxyaquic Eutrudepts on transect MC3 [*USDA-NRCS*, 1998]. All soils are shallow, loamy-skeletal with high mica contents. The organic layer consisted of Oi, Oe, and frequently also Oa horizons and had a thickness between 2 and 43 cm [mean of 16 cm; *Wilcke et al.*, 2002]. The thickness increased with increasing altitude giving Histosols (mainly Terric Haplosaprists) above c. 2100 m. Selected soil properties are summarized in Table A-1; data were taken from the work of *Yasin* (2001). The underlying bedrock consists of interbedding of paleozoic phyllites, quartzites and metasandstones (the "Chiguinda unit" of the "Zamora series" in the work of *Hungerbühler* [1997]).

MCs 2 and 3 are entirely forested, whereas the upper part of MC 1 has been used for agriculture until about 10 years ago. This part is currently undergoing natural succession and is covered by grass and shrubs. The study forest can be classified as "bosque siempreverde montaño" (evergreen montane forest [*Balslev and Ollgaard*, 2002]) or as Lower Montane Forest [*Bruijnzeel and Hamilton*, 2000]. More information on the composition of the forest can be found in the work of *Homeier* [2004].

3.2 Field sampling

Water samples were collected between April 1998 and April 2003. Each gauging station for incident precipitation consisted of five samplers. Solution sampled by rainfall collectors was "bulk precipitation" [*Whitehead and Feth*, 1964], since collectors were open to dry deposition between rainfall events [*Parker*, 1983]. However, the contribution of dry deposition to our rainfall collectors only comprised the soluble part of particles that are large enough to sediment gravitationally. We assumed that this coarse particulate deposition was small compared with the far larger aerosol trapping capacity of the forest canopy collecting also gaseous and fine particulate deposition by impaction and therefore neglected [*Parker*, 1983]. Each of the five transects was equipped with five throughfall collectors (in May 2000, three more collectors were added on each transect). All throughfall samplers had a fixed position that was arbitrarily chosen and evenly distributed along the transects. To rove samplers after each sample collection, as suggested by *Lloyd & Marques* [1988], to improve the representativity of the sample would have resulted in an unacceptable damage to the study forest that was only accessible on very steep machete-cleared and rope-secured paths. More information on our throughfall measurement is reported in the work of *Fleischbein et al.* [2005].

At the lowermost transects of all catchments (MC1, MC2.1, MC3) five trees were equipped with stemflow collectors. Each of the throughfall and stemflow samples were combined to one sample per transect in the field. Stream water samples were weekly taken from the center of the streams at the outlet of each catchment.

Throughfall and rainfall collectors consisted of fixed 1-l polyethylene sampling bottles and circular funnels with a diameter of 115 mm. The opening of the funnel was at 0.3 m height above the soil. The collectors were equipped with table tennis balls to reduce evaporation. Incident rainfall collectors were additionally wrapped with aluminum foil to reduce the impact of radiation. Stemflow collectors were made of polyurethane foam and connected with plastic tubes to a 10-l container [*Likens and Eaton*, 1970]. In each catchment, four trees of the uppermost canopy layer and one tree fern belonging to the second tree layer were used for stemflow measurements. The species were selected to be representative of the study forest although this was difficult because of its high plant diversity. A list of the selected species is given by *Fleischbein et al.* [2005].

3.3 Hydrological measurements

Rainfall, throughfall and stemflow were measured weekly by recording single volumes for each collector. Additionally each catchment was equipped with a tipping bucket rain gauge (NovaLynx® 260-2500, NovaLynx Corporation, Grass Valley, U.S.A.) to obtain higher resolution data of throughfall volume. Due to logger breakdowns and funnel blockings the data set was incomplete. Missing data was substituted by regression of precipitation data (RF) of the meteorological station on throughfall (*TF*, equations (1)-(3))

$$\text{Throughfall}_{MC1} = 0.651 \text{ Rainfall}_{metstation} + 2.483 \quad (r^2=0.83, n=263; 16/4/1998\text{-}14/5/2003) \tag{1}$$

$$\text{Throughfall}_{MC2} = 0.707 \text{ Rainfall}_{metstation} + 2.583 \quad (r^2=0.88, n=263; 16/4/1998\text{-}14/5/2003) \tag{2}$$

$$\text{Throughfall}_{MC3} = 0.831 \text{ Rainfall}_{metstation} + 0.2913 \quad (r^2=0.86, n=263; 15/4/1998\text{-}14/5/2003) \tag{3}$$

To quantify surface flow, in April 1998 Thompson (V-notch) weirs (90°) with sediment basins were installed in the lower part of each catchment and water level was instantaneously recorded hourly with a pressure gauge (water level sensor). Additionally, water level was measured by hand after sampling of stream water. The empirical equations (4)-(6) were used to convert the water level h [cm] to surface flow q [l s^{-1}].

$$q(MC1) = 0.0140 \ h^{2.5156} \tag{4}$$
$$q(MC2) = 0.0146 \ h^{2.5575} \tag{5}$$
$$q(MC3) = 0.0081 \ h^{2.7067} \tag{6}$$

Equations (4)-(6) were derived from direct measurement of the surface flow at different water levels (MC1: n=31; MC2: n = 28; MC3: n= 24 direct surface flow measurements). Unfortunately, logger breakdowns occurred during the runoff measurement likely because of the frequently wet conditions in the studied forest. Data gaps were closed by means of the hydrological modeling program TOPMODEL [*Beven et al.*, 1995] as described by *Fleischbein et al.* [2006].

3.4 Chemical analyses

In the solution samples, Cl^- concentrations were determined with a Cl^--specific ion electrode (Orion® 9617 BN) immediately after collection in Ecuador during the first three years. After export of the filtered 100-ml aliquots from Ecuador to Germany in frozen state, Ca, Mg, K, and Na concentrations were determined with flame atomic absorption spectroscopy (AAS). In the fourth to fifth year Cl^- was also analyzed with with a segmented continuous flow analyzer (CFA San plus, Skalar®). Ca, Mg, and K are referred to as "base metals" in the following. Na is used to assess the potential marine influence on the chemical composition of rainwater. Furthermore, water samples were analyzed colorimetrically with the CFA for concentrations of dissolved inorganic nitrogen (NH_4-N and NO_3-N; NO_3-N contained a small unquantified contribution of NO_2-N), and total dissolved nitrogen (TDN) concentrations (after UV oxidation to NO_3). Additionally, total dissolved sulfur (TDS) concentrations were determined by inductively coupled plasma optical emission spectrometry (ICP-OES, Integra XMP, GBC Scientific Equipment®). Concentrations of N species and sulfur were only used as markers for biomass burning-related input from the atmosphere.

3.5 Calculation of deposition rates and element fluxes

Annual element fluxes were calculated for rainfall, throughfall, and stemflow by multiplying the respective annual volume-weighted mean (VWM) concentrations with the annual water fluxes. Element fluxes with surface runoff were calculated by multiplying flow-weighted mean (FWM) concentrations with the measured or modeled annual surface runoff and referring the annual flux to the surface area of the catchments (MC1: 8 ha, MC2: 9.1 ha, MC3: 13 ha). To estimate the dry deposition and quantify canopy leaching we used the model of *Ulrich* [1983]. The total deposition (*TD*) of an element *i* was calculated with equation (7).

$$TD_i = RD_i + DD_i \qquad (7)$$

Here, RD is bulk rainfall deposition measured at the three gauging stations adjacent to the study forest (including water-soluble coarse particulate deposition) and DD is dry deposition estimated with equation (8). This estimate of DD includes water-soluble dry particulate and gaseous deposition.

$$DD_i = [(TFD_{Cl} + SFD_{Cl})/RD_{Cl}] RD_i - RD_i \qquad (8)$$

where TFD_{Cl} represents the throughfall deposition of Cl⁻ and SFD_{Cl} the stemflow deposition of Cl⁻. The quotient $(TFD_{Cl} + SFD_{Cl})/RD_{Cl}$ is called deposition ratio. The canopy budget (LEA) was calculated with equation (9). Positive values of LEA indicate leaching, negative ones uptake of an element i by the canopy.

$$LEA_i = (TFD_i + SFD_i) - RD_i - DD_i \qquad (9)$$

To test whether the interannual variation in base metal deposition was related to biomass burning we performed a principal component analysis after varimax rotation (SPSS® 13.0 for windows®) for Ca, Mg, K, N species, and S. If elements loaded the same principal component we considered them to have the same or collocated sources.

3.6 Remote sensing and transport pathway reconstruction

To compare tropospheric dust emission of the Sahara to atmospheric input to Andean montane rain forest we used the aerosol index of the "NASA Ozone Processing Team" based on measured radiances recorded by the Total Ozone Mapping Spectrometer (TOMS [*Torres et al.*, 2002]) on board the NASA satellite "EARTH Probe". We used weekly and monthly averages of daily TOMS aerosol index (AI) data. For reconstruction of transport pathways of air to our study site we calculated 13-day backward trajectories (target heights: 3000, 4000, and 6000m. a. s. l., all under the average cloud top heights at the study site [*Bendix et al.*, 2006], using the Hybrid Single Particle Lagrangian Integrated Trajectories (HYSPLIT) model [*Draxler and Hess*, 1998] provided by the NOAA Air Recources Laboratory (ARL). To validate backward trajectories crossing the intertropical convergence zone we additionally calculated 13 day forward trajectories from western African sources (30°N, 9°W; 23°N, 15°W; 15°N, 16°W).

To assess dust transport conditions across Amazonia we evaluated a typical Saharan dust transport corridor representing >85% of noon trajectory passage at 3000, 4000, and 6000m target heights across Amazonia by backward trajectory analysis (HYSPLIT). This corridor was split into three zones, each roughly representing the distance across which Saharan dust can be transported in a single day (Figure B-1). We calculated a daily mean precipitation based on the data of 5-12 meteorological stations in each zone run by the Instituto Nacional de Pesquisas Espaciais ([INPE]; ~300,000 measurements, data with obvious errors [<1%] removed, data are available at http://tempo.cptec.inpe.br:9080/PCD). We considered the transport conditions for dust as optimal, if Saharan air was able to cross a zone without being exposed to precipitation (=0 mm mean daily precipitation of all meteorological stations in the considered zone). Furthermore, we assumed that this had to be the case on three consecutive days from zone 1 in the east to zone 3 in the west. We took the number of such days per month (counted as date of the trajectory's arrival at our study site) as an index of the transport probability of Saharan dust over Amazonia.

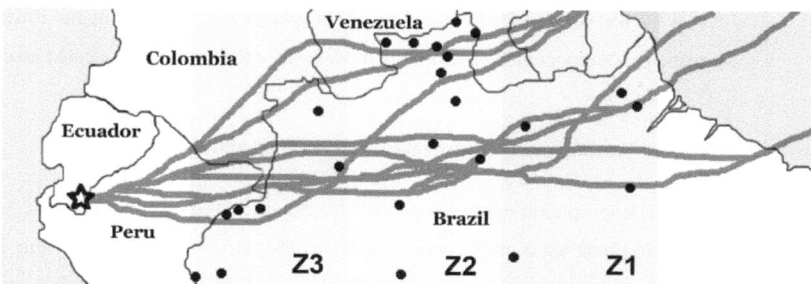

Figure B-1: Sahara dust transport corridor across Amazonia. Location of the 24 meteorological stations of INPE (solid circles), the three defined passage zones (shaded, Z1-Z3) and the study site (open star). As an example, daily HYSPLIT-backward trajectories (3000, 4000, and 6000m target heights, coming in from the western Sahara) are drawn as grey lines for the week of 01-07 May 1999. The transported aerosol experienced good transport probability across Amazonia (Figure B-3) and was washed out at our study site (126 mm precipitation in that week [01-07/May/1999]). There was no biomass burning in the transport corridor (as detected by the satellite NOAA-12). The Ca total deposition at the Ecuadorian montane forest for that week (01-07 May 1999) was 3.6 kg ha^{-1}.

4. Results and Discussion

4.1 Base metal input and response of the ecosystem

Calcium. We observed a strong interannual variation in Ca deposition (Table B-1, Figure B-2). This variation in Ca deposition changed the Ca budget of the investigated catchments from overall loss during low-input years to overall retention during high-input years, indicating a direct response of the ecosystem to Ca deposition (Figure B-2). One possible explanation of the retention of a nutrient in the ecosystem is the shortage in supply of the retained nutrient to the vegetation. Retention of Ca was observed at our study sites from 1999 to 2002. Interestingly, the most pronounced accumulation of Ca was observed in the hydrological year 1999/2000, which is also the year of the highest Ca input (Figure B-2). The surprisingly stable Ca export via streamflow during most monitored years suggests an unavoidable Ca output by surface flow of around 5 kg ha^{-1} a^{-1} (Figure B-2). This unavoidable Ca loss might be a consequence of the law of electroneutrality because the leaching of anions (chloride, nitrate, organic anions etc.) necessarily requires simultaneous cation leaching, including Ca [*Perakis et al.*, 2006]. There is also a contribution to Ca export by weathering of parent rock in the deep subsoil which might not be accessible by plants [*Hedin et al.*, 2003]. A rough estimate based on the gradient in Ca concentrations between soil water at 30 cm depth in the mineral soil (after passing more than 90% of the rooting zone [*Soethe et al.*, 2006]) and groundwater-fed surface flow suggests a weathering rate of 2.5 kg Ca ha^{-1} a^{-1}. In 1999/2000, the Ca output with surface flow was markedly higher than in all other years, because part of the deposited Ca was lost. This loss is related to Ca export via fast near-surface flow in soil carrying high nutrient loads quickly to the stream during rainstorm events. The occurrence of fast near-surface flow in response to rain storms was shown by *Goller et al.*, [2005]. We estimate that the near-surface flow accounted for a Ca export of 6 kg ha^{-1} a^{-1} in 1999/2000. This estimate is derived by subtracting the long-term Ca output as result of subsoil weathering from the observed total Ca output in 1999/2000. During the other monitored years only 1-2 kg Ca ha^{-1} a^{-1} was exported via this pathway. Nevertheless, about 15 kg of the deposited Ca per hectare were retained in the forest in 1999/2000. The retention of 'additional' deposited Ca between 1999 and 2002 might indicate that Ca enhanced plant growth, possibly by altering N uptake [*Hedin et al.*, 2003; *McLaughlin and Wimmer*, 1999], and would imply that nutrient accretion because of accelerated plant growth increased.

Table B-1: Mean annual fluxes of Ca, Mg, and K [kg ha^{-1} a^{-1}] and standard deviations of rainfall, throughfall, dry deposition and canopy budget of the five hydrological years of the three microcatchments (MC1- MC3).

		1998/1999			1999/2000			2000/2001			2001/2002			2002/2003		
	Units	Ca	Mg	K	Ca	Mg	K	Ca	Mg	K	Ca	Mg	K	Ca	Mg	K
rainfall	[kg ha^{-1} yr^{-1}]	3.4	1.2	3.7	16	6.7	16	3.5	1.7	4.5	3.0	1.3	7.2	2.7	1.0	8.0
		-	-	-	-	-	-	±0.86	±0.5	±0.7	±0.47	±0.3	±2.4	±0.29	±0.2	±1.5
throughfall	[kg ha^{-1} yr^{-1}]	19	11	122	27	14	126	19	12	135	18	11	120	21	14	147
		±6.5	±6.6	±32	±12	±11	±82	±16	±12	±87	±12	±8.2	±47	±17	±12	±71
stemflow	[kg ha^{-1} yr^{-1}]	0.44	0.21	3.1	0.54	0.22	2.3	0.30	0.15	2.3	0.29	0.15	2.3	0.29	0.15	2.2
		±0.21	±0.11	±1.1	±0.21	±0.10	±0.74	±0.16	±0.08	±0.58	±0.13	±0.06	±0.39	±0.14	±0.07	±0.25
dry deposition	[kg ha^{-1} yr^{-1}]	2.8	1.0	3.0	12	5.2	12	6.5	3.2	8.7	4.7	2.0	11	1.4	0.5	4.4
		-	-	-	-	-	-	±1.9	±1.3	±3.9	±0.44	±0.27	±1.9	±0.64	±0.28	±2.4
canopy budget	[kg ha^{-1} yr^{-1}]	9.8	5.8	108	-6.2	-3.1	66	14	11	159	15	11	123	24	16	162
		-	-	-	-	-	-	±23	±16	±101	±13	±9.5	±41	±20	±14	±76

Magnesium. The temporal course of Mg deposition during the monitoring period strongly resembled that of Ca (r^2=0.85 for concentrations and fluxes; n=263 weeks, Figures B-2, B-3). Retention of Mg in the catchments was observed during the high input hydrological year of 1999/2000 and during the following year (6 and 0.9 kg ha^{-1}, respectively). Considering the higher availability of Mg than of Ca in soil (Table A-1) and an average (molar) Ca/Mg ratio of 1.5 in leaves (n=184) and 1.9 (n=402) for total standing biomass we interpret the Mg retention in the system as an indirect effect of the assumed altered nutrient accretion induced by Ca deposition. The role of six-fold enhanced Mg deposition certainly would be different in Amazonian forests where Mg is the limiting factor.

Potassium. K deposition also showed interannual variation. However, K concentrations and fluxes with rainfall were not correlated with those of Ca and Mg. During weeks with highest Ca and Mg deposition also enhanced K deposition occurred (results not shown), but K deposition was also elevated in the hydrological year of 2001/2002 (Figure B-2). K deposition was low 9.8-30 kg ha^{-1} a^{-1} compared with K leaching from the canopy 66-162 kg ha^{-1} a^{-1}, which was different to Ca and Mg where deposition (4.4-29 and 1.6-12 kg ha^{-1} a^{-1}, respectively) was similar to canopy uptake or leaching (-6.2-24 and -3.1-16 kg ha^{-1} a^{-1}, respectively). In 1999/2000, the high-input year, Ca and Mg were even retained in the canopy suggesting plant uptake, while K was still leached during this period. On balance, K was retained in the catchments in all years probably because deposited (and leached) K was specifically fixed in the peripherally widened interlayers of the abundant illites present in the soils of the study region where the sorption of the K^+ ion with its perfectly fitting ion radius results in recontraction of the interlayers [*Schrumpf et al.*, 2001].

Our results demonstrate that the variation in base metal input by deposition from the atmosphere among different years has a measurable effect on the nutrient budget of a tropical montane rain forest. Given its least availability of all base metals (on a molar basis) and its strongest retention in the studied catchments we suggest that Ca played the key role for the observed effects of base metal deposition to the monitored forest.

4.2 Source identification

To explain the interannual variation in base metal deposition from the atmosphere we tested (i) local dust, (ii) anthropogenic emissions, (iii) volcanism, (iv) biomass burning, (v) biogenic aerosols, and (vi) long-range transport as possible sources. The only possible source of local base metal-containing dust originates in the dry inner-Andean valley to the west of our study site. This is typically downwind for our study site where easterly winds dominate. Western base metal sources are furthermore shielded from our study site by a >3000m high mountain crest. Consequently, elevated base metal concentrations in rainfall (>0.5 mg l^{-1} Ca; >0.25 mg l^{-1} Mg) almost never occurred (<5%) during the rare periods of predominant westerly winds (ground-based measurements on the ridge [*Fabian et al.*, 2005]). Consequently, wind systems passing by the Atacama Desert (as well arriving at the study site from westerly direction as indicated by HYSPLIT trajectories) did also not cause elevated base metal concentrations. Rainfall samples collected during westerly wind-dominated periods had a molar Na/Cl ratio of 1.3. This was closer to the maritime molar Na/Cl ratio of 0.86 than in rainfall samples collected during periods of elevated base metal deposition and indicated a contribution of sea salt from the Pacific to our study region. During easterly wind-dominated periods with elevated base metal deposition the molar Na/Cl ratio was 3.2. Thus, the Pacific contribution was not associated with elevated base metal deposition. All likely sources of anthropogenic emissions in the short and midrange transport distance are also situated to the W and can be excluded as a source for the reasons mentioned above. We did not observe a relation between the reported volcanic eruptions during the monitored period and the base metal concentration in our weekly rainfall samples. Neither of the two major eruptions during the monitored period produced a distinct peak in Ca concentrations in rainfall (Figure B-3). This was also the case for Mg and K (not shown).

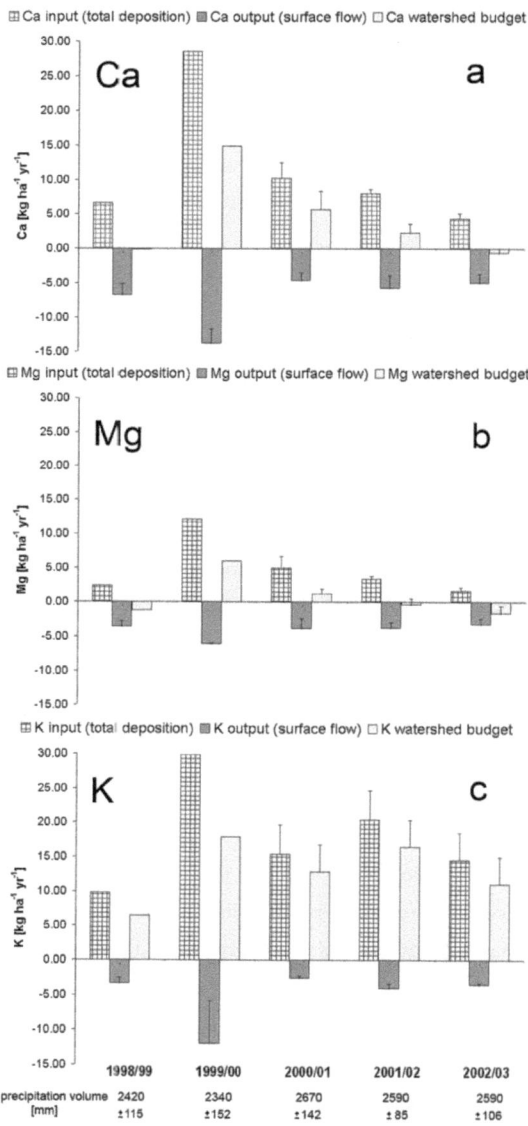

Figure B-2: Mean annual input by deposition, output by surface flow and microcatchment budget of (a) Ca, (b) Mg and (c) K of three microcatchments of tropical montane rain forest in south Ecuador between May 1998 and April 2003. Negative values of microcatchment budget indicate loss, positive values accumulation of base metals in the ecosystem. Error bars indicate standard deviations among the three catchments.

The Amazonian burning season peaks each year in August/September [INPE, http://www.cptec.inpe.br/queimadas] and was therefore not in phase with elevated Ca deposition peaking in May/July (Figure B-3). Furthermore, a varimax-rotated principal component analysis of the concentrations of Ca, Mg, K, Na, chloride, nitrate, ammonium, organic N, and S in rainfall indicated that Ca, Mg, and K had a different origin than the typical fire markers S and N, because Ca, Mg, and K highly loaded another principal component. Another probable source of Ca, Mg, and K might be primary biogenic aerosols (spores, pollen, plant parts etc.) which form a substantial part of Amazonian aerosol emissions [*Graham et al.*, 2003; *Mahowald et al.*, 2005]. Although primary biogenic aerosols might contribute significantly to base cation deposition our study site they cannot explain interannual variations as they are usually emitted at a constant rate [*Mahowald et al.*, 2005]. However, *Roberts et al.* [2001] found Ca in the mineral and not in the biogenic fraction when analyzing aerosols over a rain forest in Congo.

Long-range transport of soil dust from the Amazon basin is not a likely source of base metals because of the low alkaline earth metal concentrations of Amazonian ecosystems on predominantly deeply weathered and nutrient-depleted soils [*Cuevas and Medina*, 1986; , 1988; *Swap et al.*, 1992].

Besides sea spray, the only significant atmospheric source of base metals for Amazonia is Saharan dust, which contains appreciable amounts of Ca, Mg, and K (up to 18wt% of Ca oxides, calcites, and dolomites, up to 4wt% of mafic Mg and 3wt% of K_2O [*Claquin et al.*, 1999; *Moreno et al.*, 2006; *Reid et al.*, 2003]). Therefore we compared the temporal course of aerosol emission activity over the Sahara as detected by TOMS with the temporal course of base metal deposition at our study site. The amount of deposited Ca and Mg corresponded well to Saharan source activity as detected by TOMS. Strong desert storms leading to high aerosol concentrations over the west Sahara, a known source for Ca- and Mg-rich aerosols [*Claquin et al.*, 1999; *Moreno et al.*, 2006; *Reid et al.*, 2003] corresponded to periods of high Ca and Mg deposition at our study site (Figure B-3 A,C,D, and H). These sources are saline dry lakes from where also halites are blown out which might explain the high Na/Cl ratios of the dust deposits [*Prospero et al.*, 2002]. Source activity in the eastern Sahara or the Sahel, both known to be substantially weaker emission sources for Ca and Mg [*Claquin et al.*, 1999; *Moreno et al.*, 2006; *Reid et al.*, 2003] caused much lower Ca and Mg depositions (Figure B-3, E,G, and I).

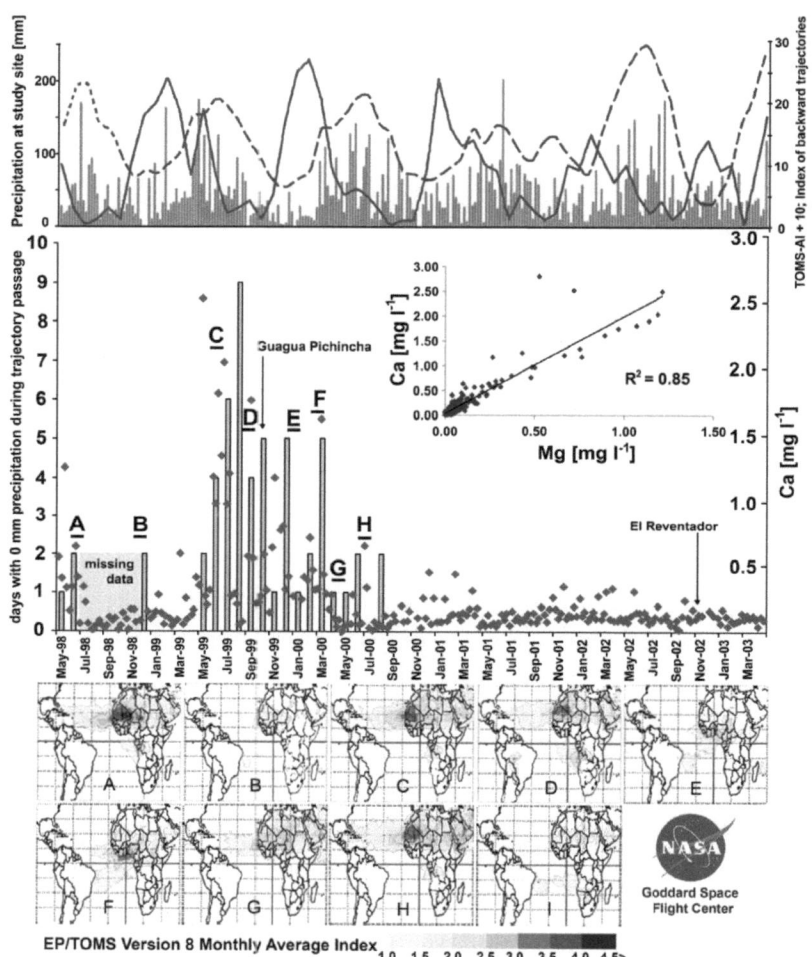

Figure B-3: Upper graph: Temporal course of precipitation at the study site (columns), the number of daily HYSPLIT backward trajectories per month which originated in the Sahara and reached the study site at noon at the target heights 3000, 4000 or 6000 m a.s.l. (solid line), and the monthly mean TOMS aerosol index (AI) of the west African source region (35.5°N-9.5°N, 17°W-11°E, broken line). Central graph: Temporal course of Ca concentrations in rainfall samples (diamonds, inlet figure illustrates the close relation between Ca and Mg deposition), and probability of Saharan dust transport across Amazonia expressed in absolutely dry conditions (i.e., number of days per month with 0 mm precipitation along the passage of daily trajectories across the Amazonian dust transport corridor [Figure B-1], columns). Two volcanic eruptions during the monitoring period are indicated by arrows and name of volcano. Both volcanoes are located around 500 km to the N or NNE, respectively. Lower graph: Monthly aerosol emissions over the Sahara as detected by TOMS (panels A-I, corresponding letters in the central graph indicate the time covering). Between July and November 1998 precipitation data coverage across Amazonia was insufficient to determine transport probability (no data for 'Z3' of Figure B-1).

When Saharan aerosol emissions were low or absent, we never observed elevated Ca or Mg concentration in bulk deposition higher than 0.5 mg l^{-1} during the period of elevated base cation deposition (Figure B-3, B). Backward trajectory analysis (HYSPLIT) revealed that the assumption of a connection between Saharan aerosol emissions and Ca and Mg deposition to our study site was reasonable for the monitored period between May 1999 and April 2003. All weekly rainfall samples with elevated concentrations (>0.5 mg l^{-1} Ca; >0.25 mg l^{-1} Mg) were linked to at least one day on which air masses from the Sahara arrived. The highest Ca and Mg deposition events were found to be under predominant influence of Saharan air transport from Ca-rich west Saharan sources (Figures B-2, B-3). The temporal course of K deposition suggested that it was also deposited during Saharan dust input periods. However, K had at least a second, independent source besides Saharan dust because K concentrations were also elevated for a prolonged period in 2002/2003, during which Saharan dust deposition was unlikely (Figure B-3).

The frequently observed seasonality of dust transport across the Atlantic ocean with a predominantly northern pathway to the Caribbean and the United States from May to July and a southern pathway to South America from November to March [e.g. *Engelstaedter et al.*, 2006; *Prospero and Lamb*, 2003; e.g. *Swap et al.*, 1992] caused little differences in aerosol transport between the north-hemispherical and the south-hemispherical part of the transport corridor relevant for this study (monthly mean TOMS AI for 4.5°N-0.5°N compared to 0.5°S-4.5°S, both 51°W-29°W, slope of the regression line= 0.97, r=0.85). During the predominantly northern pathway-season we found back- and forward trajectories crossing the intertropical convergence zone from western Saharan source regions towards our study site. These findings indicate that our study site might receive dust deposition during both transport seasons, although the impact of dust transported via the Caribbean pathway further to the north appears to be reasonably weakened. Satellite observations support this finding by detection of Saharan dust transport to Amazonia and beyond for both dust transport seasons [*Kaufman et al.*, 2005].

4.3 Relation of precipitation and Saharan dust transport across Amazonia

The assumption that Saharan dust transport across Amazonia was only possible during restricted periods was confirmed by our observations. There were periods of substantial dust outbreaks in the Sahara with no enhanced base metal deposition at our study site. The elevated Ca deposition to north Andean forests in 1999/2000 coincided with the La Niña episode at the same time. This contradicted our hypothesis that the warm ENSO period (El Niño) favors dust transport across Amazonia. The annual precipitation in the Amazonian transport corridor (Figure B-1) did not vary much between the five hydrological years (range: 2230-2660 mm a^{-1}). With a precipitation of 2470 mm a^{-1}, the hydrological year of 1999/2000 with its high base metal deposition was average. However, the number of days with optimal transport conditions for dust in the Amazonian atmosphere was different among the monitored hydrological years. Days with optimal transport conditions mainly occurred during 1999/2000 and implied a difference in rainfall distribution among the years. During "La Niña", strong rain events were interrupted by dry spells allowing for dust transport. During periods of climate conditions not affected by ENSO, precipitation was more equally distributed (Figure B-3), although data suggest a slightly elevated "background" cation deposition when strong aerosol emission activity in the Sahara met high numbers of trajectories to the study side (e.g. between January to April 1998, December 2000 to April 2001, and November 2001-May 2002; Figure B-3). This might indicate that a small part of desert dust might be able to cross a wet Amazon basin.

More than 95% of weekly samples with elevated Ca and Mg deposition (>0.5 mg l^{-1} Ca; >0.25 mg l^{-1} Mg) in the Ecuadorian forest were collected during periods of optimal transport conditions (Figure B-3). This corresponds to the finding that high Saharan dust deposition to Amazonia is following rainfalls of major wet systems [*Swap et al.*, 1992]. Therefore, we hypothesize that Saharan dust transport across Amazonia (and thus Ca supply) to Ecuadorian montane forests are linked to the ENSO cycle via the resulting changes of precipitation patterns in the region.

5. Conclusions

Our results demonstrate that the Amazon basin does not necessarily act as the final sink for Saharan dust in spite of its high precipitation. We conclude that tropical montane rain forest in South Ecuador receives substantial contributions of base metals from north African sources. This base metal deposition was large enough to change the overall Ca and Mg budget in the north Andean montane forests. To our knowledge, this is the first direct confirmation at the ecosystem level of the hypothesis that Amazonian forest nutrition is connected to dust transport from the Sahara [*Kaufman et al.*, 2005; *Reichholf*, 1986; *Swap et al.*, 1992].

We found a coincidence between ENSO and forest nutrition. The frequency of ENSO events has been proposed to increase because of global warming [*Timmermann et al.*, 1999]. Thus the nutrient budget of our study forest could be shifted by more frequent inputs of desert dust to an unknown state.

6. Acknowledgements

We thank C. Valarezo for his support in Ecuador, M. O. Andreae, S. Foley, R. Rollenbeck and H. Wernli for discussion, K. Fleischbein, R. Goller, M. Sequeira, S. Yasin and numerous student helpers for help with data acquisition, the NASA for providing the TOMS data and graphs, the NOAA for admission of HYSPLIT data, and INPE-CPTEC for the meteorological data of Amazonia. This work was funded by the German Research Foundation (DFG FOR 402, Wi1601/5-1,-2,-3).

7. References

Balslev, H., and B. Ollgaard (2002), Mapa de vegetación del sur de Ecuador, in *Botánica Austroecuatoriana. Estudios sobre los recursos vegetales en las provincias de El Oro, Loja y Zamora-Chinchipe*, edited by M. Z. Aguirre, et al., pp. 51-64, Ediciones Abya-Yala, Quito, Ecuador.

Bendix, J., R. Rollenbeck, D. Gottlicher, and J. Cermak (2006), Cloud occurrence and cloud properties in Ecuador, *Climate Research*, 30(2), 133-147.

Beven, K. J., R. Lamb, P. F. Quinn, and R. B. Romanowicz (1995), TOPMODEL, in *Computer Models of Watershed Hydrology*, edited by V. P. Singh, pp. 627-668, Water Resource Publications, Colorado.

Brooks, T. M., et al. (2002), Habitat loss and extinction in the hotspots of biodiversity, *Conservation Biology*, 16(4), 909-923.

Bruijnzeel, L. A. (1991), Nutrient input-output budgets for tropical forest ecosystems: a review, *Journal of Tropical Ecology*, 7, 1-24.

Bruijnzeel, L. A., and L. S. Hamilton (2000), Decision time for cloud forests, 1-41 pp, IHP-Unesco and WWF International, Paris, Amsterdam.

Campo, J., J. M. Maass, V. J. Jaramillo, and A. M. Yrizar (2000), Calcium, potassium, and magnesium cycling in a Mexican tropical dry forest ecosystem, *Biogeochemistry*, 49(1), 21-36.

Chiapello, I., C. Moulin, and J. M. Prospero (2005), Understanding the long-term variability of African dust transport across the Atlantic as recorded in both Barbados surface concentrations and large-scale Total Ozone Mapping Spectrometer (TOMS) optical thickness, *Journal of Geophysical Research-Atmospheres*, 110(D18), D18S10, doi:10,1029/2004JD005132.

Claquin, T., M. Schulz, and Y. J. Balkanski (1999), Modeling the mineralogy of atmospheric dust sources, *Journal of Geophysical Research-Atmospheres*, *104*(D18), 22243-22256.

Cuevas, E., and E. Medina (1986), Nutrient dynamics within Amazonian forests I. Nutrient flux in fine litter fall and efficiency of nutrient utilization, *Oecologia*, *68*, 466-472.

Cuevas, E., and E. Medina (1988), Nutrient dynamics within amazonian forests II. Fine root growth, nutrient availability and litter decomposition, *Oecologia*, *76*, 222-235.

Draxler, R. R., and G. D. Hess (1998), An overview of the HYSPLIT_4 modelling system for trajectories, dispersion and deposition, *Australian Meteorological Magazine*, *47*(4), 295-308.

Engelstaedter, S., I. Tegen, and R. Washington (2006), North African dust emissions and transport, *Earth-Science Reviews*, *79*(1-2), 73-100.

Fabian, P., M. Kohlpaintner, and R. Rollenbeck (2005), Biomass burning in the Amazon-fertilizer for the mountaineous rain forest in Ecuador, *Environmental Science and Pollution Research*, *12*(5), 290-296.

Fleischbein, K., W. Wilcke, R. Goller, J. Boy, C. Valarezo, W. Zech, and K. Knoblich (2005), Rainfall interception in a lower montane forest in Ecuador: effects of canopy properties, *Hydrological Processes*, *19*(7), 1355-1371.

Fleischbein, K., W. Wilcke, C. Valarezo, W. Zech, and K. Knoblich (2006), Water budgets of three small catchments under montane forest in Ecuador: experimental and modelling approach, *Hydrological Processes*, *20*(12), 2491-2507.

Formenti, P., et al. (2001), Saharan dust in Brazil and Suriname during the Large-Scale Biosphere-Atmosphere Experiment in Amazonia (LBA) - Cooperative LBA Regional Experiment (CLAIRE) in March 1998, *Journal of Geophysical Research-Atmospheres*, *106*(D14), 14919-14934.

Goller, R., W. Wilcke, M. J. Leng, H. J. Tobschall, K. Wagner, C. Valarezo, and W. Zech (2005), Tracing water paths through small catchments under a tropical montane rain forest in south Ecuador by an oxygen isotope approach, *Journal of Hydrology*, *308*(1-4), 67-80.

Graham, B., et al. (2003), Composition and diurnal variability of the natural Amazonian aerosol, *Journal of Geophysical Research-Atmospheres*, *108*(D24), 4765, doi:4710,1029/2003JD004049.

Hamilton, L. S., J. O. Juvik, and F. N. Scatena (1995), *Tropical Montane Cloud Forests*, Springer, New York.

Hedin, L. O., P. M. Vitousek, and P. A. Matson (2003), Nutrient losses over four million years of tropical forest development, *Ecology*, *84*(9), 2231-2255.

Homeier, J. (2004), Baumdiversität, Waldstruktur und Wachstumsdynamik zweier tropischer Bergregenwälder in Ecuador und Costa Rica, Ph. D thesis, University of Bielefeld, Bielefeld, Germany.

Hungerbühler, D. (1997), Neogene basins in the Andes of southern Ecuador: evolution, deformation and regional tectonic implications, Ph. D. thesis, ETH, Zürich.

Kaufman, Y. J., I. Koren, L. A. Remer, D. Tanre, P. Ginoux, and S. Fan (2005), Dust transport and deposition observed from the Terra-Moderate Resolution Imaging Spectroradiometer (MODIS) spacecraft over the Atlantic ocean, *Journal of Geophysical Research-Atmospheres*, *110*(D10), D10S12, doi:10,1029/2003JD004436.

Likens, G. E., and J. S. Eaton (1970), A polyurethane stemflow collector for trees and shrubs, *Ecology*, *51*, 938-939.

Likens, G. E., C. T. Driscoll, D. C. Buso, T. G. Siccama, C. E. Johnson, G. M. Lovett, D. F. Ryan, T. Fahey, and W. A. Reiners (1994), The Biogeochemistry of Potassium at Hubbard Brook, *Biogeochemistry*, *25*(2), 61-125.

Likens, G. E., et al. (1998), The biogeochemistry of calcium at Hubbard Brook, *Biogeochemistry*, *41*(2), 89-173.

Lloyd, C. R., and A. Marques (1988), Spatial variability of throughfall and stemflow measurements in Amazonian rain forest, *Agricultural and Forest Meteorology*, *47*, 63-73.

Mahowald, N. M., P. Artaxo, A. R. Baker, T. D. Jickells, G. S. Okin, J. T. Randerson, and A. R. Townsend (2005), Impacts of biomass burning emissions and land use change on Amazonian atmospheric phosphorus cycling and deposition, *Global Biogeochemical Cycles*, *19*(4), GB4030, doi:4010,1029/2005GB002541.

McLaughlin, S. B., and R. Wimmer (1999), Tansley Review No. 104 - Calcium physiology and terrestrial ecosystem processes, *New Phytologist*, *142*(3), 373-417.

Moreno, T., X. Querol, S. Castillo, A. Alastuey, E. Cuevas, L. Herrmann, M. Mounkaila, J. Elvira, and W. Gibbons (2006), Geochemical variations in aeolian mineral particles from the Sahara-Sahel Dust Corridor, *Chemosphere*, *65*(2), 261-270.

Okin, G. S., N. Mahowald, O. A. Chadwick, and P. Artaxo (2004), Impact of desert dust on the biogeochemistry of phosphorus in terrestrial ecosystems, *Global Biogeochemical Cycles*, *18*(2), GB2005, doi:2010,1029/2003GB002145.

Parker, G. G. (1983), Throughfall and stemflow in the forest nutrition cycle, *Advances in Ecological Research*, *13*, 57-133.

Perakis, S., D. Maguire, T. Bullen, K. Cromack, R. Waring, and J. Boyle (2006), Coupled nitrogen and calcium cycles in forests of the Oregon Coast Range, *Ecosystems*, *9*(1), 63-74.

Prospero, J. M., P. Ginoux, O. Torres, S. E. Nicholson, and T. E. Gill (2002), Environmental characterization of global sources of atmospheric soil dust identified with the Nimbus 7 Total Ozone Mapping Spectrometer (TOMS) absorbing aerosol product, *Reviews of Geophysics*, *40*(1), 1002, doi:1010.1029/2000RG000095.

Prospero, J. M., and P. J. Lamb (2003), African droughts and dust transport to the Caribbean: Climate change implications, *Science*, *302*(5647), 1024-1027.

Reichholf, J. H. (1986), Is saharan dust a major source of nutrients for the amazonian rainforest *Studies on Neotropical Fauna and Environment 21*, 251-255.

Reid, E. A., J. S. Reid, M. M. Meier, M. R. Dunlap, S. S. Cliff, A. Broumas, K. Perry, and H. Maring (2003), Characterization of African dust transported to Puerto Rico by individual particle and size segregated bulk analysis, *Journal of Geophysical Research-Atmospheres*, *108*(D19), 8591, doi:8510,1029/2002JD002935.

Richter, M. (2003), Using epiphytes and soil temperatures for eco-climatic interpretations in Southern Ecuador, *Erdkunde*, *57*, 161-181.

Roberts, G. C., M. O. Andreae, W. Maenhaut, and M. T. Fernandez-Jimenez (2001), Composition and sources of aerosol in a central African rain forest during the dry season, *Journal of Geophysical Research-Atmospheres*, *106*(D13), 14423-14434.

Schrumpf, M., G. Guggenberger, C. Schubert, C. Valarezo, and W. Zech (2001), Tropical montane forest soils: development and nutrient status along an altitudinal gradient in the south Ecuadorian Andes, *Die Erde*, *132*, 43-59.

Schütz, L., R. Jaenicke, and H. Pietrek (1981), Saharan dust transport over the North Atlantic Ocean, *Geological Society of America, special paper 186*, 87-100.

Shaul, O. (2002), Magnesium transport and function in plants: the tip of the iceberg, *Biometals*, *15*(3), 309-323.

Soethe, N., J. Lehmann, and C. Engels (2006), The vertical pattern of rooting and nutrient uptake at different altitudes of a south ecuadorian montane forest, *Plant and Soil*, *286*(1-2), 287-299.

Swap, R., S. Garstang, S. Greco, R. Talbot, and P. Kallberg (1992), Saharan dust in the Amazon basin, *Tellus*, *44*(B), 133-149.

Timmermann, A., J. Oberhuber, A. Bacher, M. Esch, M. Latif, and E. Roeckner (1999), Increased El Nino frequency in a climate model forced by future greenhouse warming, *Nature*, *398*(6729), 694-697.

Torres, O., P. K. Bhartia, J. R. Herman, A. Sinyuk, P. Ginoux, and B. Holben (2002), A long-term record of aerosol optical depth from TOMS observations and comparison to AERONET measurements, *Journal of the Atmospheric Sciences*, *59*(3), 398-413.

Tripler, C. E., S. S. Kaushal, G. E. Likens, and M. T. Walter (2006), Patterns in potassium dynamics in forest ecosystems, *Ecology Letters*, *9*(4), 451-466.

Ulrich, B. (1983), *Effects of Accumulation of Air Pollutants in Forest Ecosystems*, Reidel, Dordrecht.

USDA-NRCS, U. S. D. o. A.-N. R. C. S. (1998), *Keys to Soil Taxonomy*, Pocahontas Press, Washington DC.

Whitehead, H. L., and J. H. Feth (1964), Chemical composition of rain, dry fallout and bulk precipitation at Menlo Park, California, *Journal of Geophysical Research*, *69*, 3319-3333.

Wilcke, W., S. Yasin, C. Valarezo, and W. Zech (2001), Change in water quality during the passage through a tropical montane rain forest in Ecuador, *Biogeochemistry*, *55*(1), 45-72.

Wilcke, W., S. Yasin, U. Abramowski, C. Valarezo, and W. Zech (2002), Nutrient storage and turnover in organic layers under tropical montane rain forest in Ecuador, *European Journal of Soil Science*, *53*(1), 15-27.

Wilcke, W., H. Valladarez, R. Stoyan, S. Yasin, C. Valarezo, and W. Zech (2003), Soil properties on a chronosequence of landslides in montane rain forest, Ecuador, *Catena*, *53*(1), 79-95.

Zeng, N. (1999), Seasonal cycle and interannual variability in the Amazon hydrologic cycle, *Journal of Geophysical Research-Atmospheres*, *104*(D8), 9097-9106.

C Amazonian biomass burning-derived acid and nutrient deposition in the north Andean montane forest of Ecuador[a]

Jens Boy[1] Rütger Rollenbeck[2], Carlos Valarezo[3] & Wolfgang Wilcke[1]

[1] Geographic Institute, Johannes Gutenberg University, 55099 Mainz, Germany
[2] Geographic Institute, Philipps University, 35032 Marburg, Germany
[3] Universidad Nacional de Loja, Area Agropecuaria y de Recursos Naturales Renovables, Programa de Agroforestería, Ciudadela Universitaria Guillermo Falconí, Loja, Ecuador

1. Abstract

We explored the influence of biomass burning in Amazonia and northeastern Latin America on N, C, P, S, K, Ca, Mg, Al, Mn, and Zn cycles of an Andean montane forest in south Ecuador exposed to the Amazon basin between May 1998 and April 2003. We assessed the response of the element budget of three microcatchments (8-13 ha) to the variations in atmospheric deposition between the intensive burning season and outside the burning season in Amazonia. There were significantly elevated H, N, and Mn depositions during biomass burning. Elevated H deposition during biomass burning caused elevated base metal loss from the canopy and the organic horizon and deteriorated already low base metal supply of the vegetation. N was only retained during biomass burning but not during non-fire conditions when deposition was much smaller. We conclude that biomass burning-related aerosol emissions in Amazonia are large enough to substantially increase element deposition at the western rim of Amazonia. Particularly the related increase of acid deposition impoverishes already base-metal scarce ecosystems. As biomass burning is most intense during El Niño situations, a shortened ENSO cycle because of global warming likely enhances the acid deposition at our study forest.

Abbreviations: MC, microcatchment; SWB, soil water balance; AAS, atomic absorption spectroscopy; CFA, continuous flow analyzer; TDN, total dissolved nitrogen; TDS, total dissolved sulphur; TDP, total dissolved phosphorus; TOC, total organic carbon; HYSPLIT, 'Hybrid Single Particle Lagrangian Integrated Trajectories; VWM, volume weighted mean; FWM, flow weighted mean; PCA, principal component analysis; BS, base saturation; ECEC, cation exchange capacity;TD, total deposition; RD, rainfall deposition; DD, dry deposition; TFD, throughfall deposition; SFD, stemflow deposition; LEA, canopy budget; LL, litter leachate; NOAA, 'National Oceanic and Atmospheric Administration'; ARL, 'Air Resources Laboratory'; GOES, 'Geostationary Environmental Satellite'; MODIS, 'Moderate Resolution Imaging Spectrometer'; INPE, 'Instituto Nacional de Pesquisas Espaciais'; GOME, 'Global Ozone Experiment'; ENSO, El Niño Southern Oscillation

2. Introduction

Biomass burning is a major source of aerosols to the atmosphere [*Andreae et cl.*, 1988]. These aerosols can contain high concentrations of C, plant nutrients (particularly N, S, P, and K), and acids [*Allen and Miguel*, 1995; *Artaxo et al.*, 2002; *Da Rocha et al.*, 2005]. Some of the aerosol species can be transported over large distances in the atmosphere [*Pereira et al.*, 1996] and deposited to distant ecosystems [*Mahowald et al.*, 2005]. Amazonia is a source region of aerosols for downwind ecosystems such as the Andean tropical montane forest in Ecuador because of forest destruction by fire and growing agricultural activity in the region [*Kauffman et al.*, 1995]. While transport [*Freitas et al.*, 2005; *Rodhe et al.*, 2002; *Tsigaridis and Kanakidou*, 2003] and sources can be monitored with the help of satellite data [*Generoso et al.*, 2003; *Ichoku and Kaufman*, 2005; *Jaegle et al.*, 2005], it is difficult to predict the complex deposition of aerosols. Aerosols can be deposited by precipitation in wet form, scavenged by the surface of the vegetation in dry particulate and gaseous form, or deposited gravitationally in dry coarse particulate form. The size of dry deposition depends on relief and vegetation cover at the target region [*Clark et al.*, 1998]. Modeling of aerosol deposition relies on detailed knowledge of micro-climatic controls, which often is not available [*Dixon and Wisniewski*, 1995]. Thus, to explore the effect of Amazonian fire-derived deposition in the Andes, direct monitoring of wet and dry deposition is required.

Several authors postulated that biomass burning has a fertilizing effect on the ecosystems receiving the resulting deposition, particularly concerning N and P [*Da Rocha et al.*, 2005; *Fabian et al.*, 2005; *Mahowald et al.*, 2005], although N limitations of Amazonian ecosystems are increasingly questioned [*Chadwick et al.*, 1999; *Martinelli et al.*, 1999; *Wardle et al.*, 2004]. This increased nutrient deposition might threaten biodiversity hotspots in the tropics [*Phoenix et al.*, 2006], including the endangered hotspot of the tropical Andes at the eastern rim of Amazonia [*Brooks et al.*, 2002].

To estimate the effect of elevated element deposition to forest ecosystems, budgets of the various ecosystem strata can be used (e.g., canopy budget or organic horizon budget [*Clark et al.*, 1998]). Comparison of separate budgets of these strata between burning and non-burning periods offer specific insight into the effects of biomass burning in Amazonia on the remote montane ecosystems of the Andes.

By monitoring all ecosystem fluxes for five consecutive years between 1998 and 2003 in weekly resolution using a whole-catchment approach in an Andean lower tropical montane rain forest in south Ecuador we tested the hypotheses that (i) biomass burning emissions of Amazonia and other sources upwind of the north-eastern trade winds are transported to and deposited at tropical montane forests in the Andes and (ii) this deposition contributes to forest nutrition, as indicated by the retention of deposited elements in the ecosystem.

3. Material and Methods

3.1 Study site

The study area is located on the eastern slope of the "Cordillera Real", the eastern Andean cordillera in south Ecuador facing the Amazon basin at 4° 00` S and 79° 05` W. We selected three 30-50° steep and 8-13 ha large microcatchments (MC1-3) under montane forest at an altitude of 1900-2200 m above sea level (a.s.l.) for our study (Figure A-1). We installed our equipment in each MC on transects, about 20 m long with an altitude range of 10 m, on the lower part of the slope at 1900-1910 m a.s.l. (transects MC1, MC2.1, and MC3). Moreover, we installed extra instrumentation at 1950-1960 (MC2.2) and 2000-2010 m a.s.l. (MC2.3). All transects were located below closed forest canopy and aligned downhill. Three unforested sites near these microcatchments were used for rainfall gauging. Gauging site 2 existed since April 1998, gauging sites 1 and 3 were built in May 2000. All catchments drain via small tributaries into the Rio San Francisco which flows into the Amazon basin.

Within the monitored period between April 1998 and April 2003 annual precipitation ranged between 2340 and 2667 mm. Additional climate data were available from a meteorological station [*Richter*, 2003] between MC 2 and 3 (Figure A-1). June tended to be the wettest month with 302 mm of precipitation on average, in contrast to 78 mm in each of November and January, the driest months. The mean temperature at 1950 m a.s.l. was 15.5 °C. The coldest month was July, with a mean temperature of 14.5 °C, the warmest November with a mean temperature of 16.6 °C.

Recent soils have developed on postglacial landslides or possibly from periglacial cover beds [*Wilcke et al.*, 2001, 2003]. Soils are Humic Eutrudepts on transect MC1, Humic Dystrudepts on transects MC2.1, MC2.2, and MC2.3, and Oxyaquic Eutrudepts on transect MC3 [*USDA-NRCS*, 1998]. All soils are shallow, loamy-skeletal with high mica contents. The organic layer consisted of Oi, Oe, and frequently also Oa horizons and had a thickness between 2 and 43 cm [mean of 16 cm; *Wilcke et al.* 2002]. The thickness increased with increasing altitude giving Histosols (mainly Terric Haplosaprists) above c. 2100 m. Selected soil properties are summarized in Table A-1; data were taken from the work of *Yasin* [2001]. The underlying bedrock consists of interbedding of paleozoic phyllites, quartzites and metasandstones (the "Chiguinda unit" of the "Zamora series" in the work of *Hungerbühler* [1997]).

MCs 2 and 3 are entirely forested, whereas the upper part of MC 1 has been used for agriculture until about 10 years ago. This part is currently undergoing natural succession and is covered by grass and shrubs. The study forest can be classified as "bosque siempreverde montaño" (evergreen montane forest [*Balslev and Øllgaard*, 2001]) or as Lower Montane Forest [*Bruijnzeel and Hamilton*, 2000]. More information on the composition of the forest can be found in the work of *Homeier* [2004].

3.2 Field sampling

Water samples were collected between April 1998 and April 2003. Each gauging station for incident precipitation consisted of five samplers. Solution sampled by rainfall collectors was "bulk precipitation" [*Whitehead and Feth*, 1964], since collectors were open to dry deposition between rainfall events [*Parker*, 1983]. However, the contribution of dry deposition to our rainfall collectors only comprised the soluble part of particles that are large enough to sediment gravitationally. We assumed that this coarse particulate deposition was small compared with the far larger aerosol trapping capacity of the forest canopy collecting also gaseous and fine particulate deposition by impaction [*Parker*, 1983]. Each of the five transects was equipped with five throughfall collectors (in May 2000, three more collectors were added on each transect). All throughfall samplers had a fixed position that was arbitrarily chosen and evenly distributed along the transects. To rove samplers after each sample collection, as suggested by *Lloyd & Marques* [1988], to improve the representativity of the sample would have resulted in an unacceptable damage to the study forest that was only accessible on very steep machete-cleared and rope-secured paths. More information on our throughfall measurement is reported by *Fleischbein et al.* [2005]. To test the chemical stability of the precipation samples during the weekly sampling interval we added six pairs of throughfall samplers. For one selected week (30/04-06/05/2008), throughfall was collected on an event-near basis (daily) in one half of the samplers and cumulatively for one week in the other half of the samplers. The solutions from each sampler were collected and analyzed separately so that we were able to compare six volume-weighted means (from the daily sampling) with six cumulative concentrations (from the weekly sampling).

We furthermore installed three collectors for litter leachate (collection of water vertically percolating through the organic layer) at lower, central, and upper positions along the transects and three suction lysimeters for soil solution sampling at each of the 0.15 m and 0.30 m depths in the mineral soil at a central position of the transect. A combined sample for each transect was produced by bulking the single samples directly in the field. Soil solution was sampled since May 2000 after equilibration of the lysimeters in the soil for four months.

At the lowermost transects of all catchments (T1, T2.1, T3) five trees were equipped with stemflow collectors. Each of the throughfall and stemflow samples were combined to one sample per transect in the field. Stream water samples were weekly taken from the center of the streams at the outlet of each catchment.

Throughfall and rainfall collectors consisted of fixed 1-l polyethylene sampling bottles and circular funnels with a diameter of 115 mm. The opening of the funnel was at 0.3 m height above the soil. The collectors were equipped with table tennis balls to reduce evaporation. Incident rainfall collectors were additionally wrapped with aluminum foil to reduce the impact of radiation. Stemflow collectors were made of polyurethane foam and connected with plastic tubes to a 10-l container [*Likens and Eaton*, 1970]. In each catchment, four trees of the uppermost canopy layer and one tree fern belonging to the second tree layer were used for stemflow measurements. The species were selected to be representative of the study forest although this was difficult because of its high plant diversity. A list of the selected species is given by *Fleischbein et al.* [2005].

3.3 Hydrological measurements

Rainfall, throughfall and stemflow were measured weekly by recording single volumes for each collector. Additionally each catchment was equipped with a tipping bucket rain gauge (NovaLynx® 260-2500, NovaLynx Corporation, Grass Valley, CA, U.S.A.) to obtain higher resolution data of throughfall volume. Due to logger breakdowns and funnel blockings the data set was incomplete. Missing data (<1% of total data) was substituted by regression of precipitation data of the meteorological station on throughfall.

Water fluxes in soil were modeled by modifying the Soil Water Balance (SWB) model [*DVWK (Deutscher Verband für Wasserwirtschaft und Kulturbau)*, 1996]. The SWB determines water fluxes out of predefined soil layers as the sum of throughfall and stemflow volume (input) minus independently determined transpiration (output) minus (or plus) change in stored water in the soil layers as calculated by the difference in water contents of the respective soil layer between two soil water measurements. For deeper soil layers, the input is the output of the overlying soil layer. We assumed direct evaporation from the soil as negligible and derived weekly transpiration rates by partitioning the annual difference between throughfall and discharge of each catchment proportionally to the weekly evapotranspiration rates as modeled by REF-ET (described in *Fleischbein et al.* [2006]). Weekly transpiration rates were furthermore split between the soil layers according to the root length densities of the respective soil layer taken from *Soethe et al.* [2006], assuming a linear relationship between water uptake of the vegetation and fine root abundance. We used soil water-content measurements logged by FDR probes in transect MC 2.1 at 0.1, 0.2, 0.3, and 0.4 m depths for all transects since differences in soil water content were little pronounced because of the overall wet environment of the study site [*Fleischbein et al.*, 2006]. Data gaps of soil water fluxes (because of lacking soil water contents) were substituted with the help of a regression model of weekly soil water fluxes on weekly throughfall volumes.

To quantify surface flow, in April 1998 Thompson (V-notch) weirs (90°) with sediment basins were installed in the lower part of each catchment and water level was instantaneously recorded hourly with a pressure gauge (water level sensor). Additionally, water level was measured by hand after sampling of stream water. Unfortunately, logger breakdowns occurred during the runoff measurement likely because of the frequently wet conditions in the studied forest. Data gaps were closed by means of the hydrological modelling program TOPMODEL [*Beven et al.*, 1995] as described by *Fleischbein et al.* [2006].

3.4 Chemical analyses

In the solution samples, Cl^- concentrations were determined with a Cl^--specific ion electrode (Orion® 9617 BN) immediately after collection in Ecuador during the first three years. In the fourth and fifth year, Cl^- was analyzed with a segmented continuous flow analyzer (CFA San plus, Skalar®). After export of the filtered 100-ml aliquots from

Ecuador to Germany in frozen state, Ca, Mg, K, and Na concentrations were determined with flame atomic absorption spectroscopy (AAS). Ca, Mg, K, and Na are referred to as "base metals" in the following. Na and Cl⁻ concentrations are used to assess the potential marine influence on the chemical composition of rain water. Furthermore, water samples were analyzed colorimetrically with a continous flow analyzer (CFA) for concentrations of dissolved inorganic nitrogen (NH_4-N and NO_3-N + NO_2-N, hereafter referred to as NO_3-N), and total dissolved nitrogen (TDN, after UV oxidation to NO_3). Additionally, total dissolved sulfur (TDS) and total dissolved phosphorus (TDP) concentrations were determined by inductively coupled plasma optical emission spectrometry (ICP-OES, Integra XMP, GBC Scientific Equipment®). Total organic carbon (TOC) concentrations were determined with an automatic TOC analyzer (TOC 5050, Shimadzu®) and Al and Mn concentrations via inductively-coupled plasma-mass spectroscopy (ICP-MS, VG PlasmaQuad PG2 Turbo Plus, VG Elemental®). Some samples had concentrations below the detection limit of the analytical methods (0.075 mg l⁻¹ for N, 0.2 mg l⁻¹ for P, 0.3 mg l⁻¹ for S, 0.001 mg l⁻¹ for Ca, Mg, K, and Na, 0.005 mg l⁻¹ for Al, and 0.002 mg l⁻¹ for Mn). For calculation purposes, values below the detection limit were set to zero (for Ca, Mg, Na, K, and TOC: <0.01%, N and Zn <1%, Al <7%, S <25%, Mn <30%, and P <45% of the values were below detection limit). Thus, our annual means underestimate the real concentration of chemical constituents and mean concentrations can be smaller than the detection limit.

3.5 Calculation of deposition rates and element fluxes

Annual element fluxes were calculated for rainfall, throughfall, and stemflow by multiplying the respective annual volume-weighted mean (VWM) concentrations with the annual water fluxes. Element fluxes with surface runoff were calculated by multiplying flow-weighted mean (FWM) concentrations with the measured annual surface runoff and referring the annual flux to the surface area of the catchments (MC1: 8 ha, MC2: 9.1 ha, MC3: 13 ha). Lacking surface flow rates because of logger problems were substituted with the help of the hydrological computer model TOPMODEL [*Fleischbein et al.*, 2006]. To estimate the dry deposition and quantify canopy leaching we used the model of *Ulrich* [1983]. The total deposition (*TD*) of an element *i* was calculated with equation (1).

$$TD_i = RD_i + DD_i \tag{1}$$

Here, *RD* is bulk rainfall deposition measured at the three gauging stations adjacent to the study forest (including water-soluble coarse particulate deposition) and *DD* is dry deposition estimated with equation (2). This estimate of *DD* includes water-soluble dry particulate and gaseous deposition.

$$DD_i = [(TFD_{Cl} + SFD_{Cl})/RD_{Cl}] RD_i - RD_i \qquad (2)$$

where TFD_{Cl} represents the throughfall deposition of Cl^- and SFD_{Cl} the stemflow deposition of Cl^-. The quotient $(TFD_{Cl} + SFD_{Cl})/RD_{Cl}$ is called deposition ratio. The canopy budget (*LEA*) was calculated with equation (3). Positive values of LEA indicate leaching, negative ones uptake of an element *i* by the canopy.

$$LEA_i = (TFD_i + SFD_i) - RD_i - DD_i \qquad (3)$$

To test whether the seasonal variation in element deposition was related to biomass burning we performed a principal component analysis after varimax rotation (SPSS® 13.0 for Windows®) for all measured elements. If elements loaded the same principal component we considered them to have the same or related sources.

3.6 Remote sensing and transport pathway reconstruction, and definition of biomass-burning season

To compare biomass burning seasons with periods not affected by biomass burning we assigned each sampling week to one of the two classes "fire" (biomass burning) and "no fire". Therefore we calculated 13-day backward trajectories (target heights: 3000, 4000, and 6000m. a. s. l., all under the average cloud top heights at the study site, [*Bendix et al.*, 2006]) using the Hybrid Single Particle Lagrangian Integrated Trajectories (HYSPLIT) model [*Draxler and Hess*, 1998] provided by the NOAA (National Oceanic and Atmospheric Administration) Air Resources Laboratory (ARL). The trajectories were combined with a fire pixel dataset by counting fire pixels on the pathway of each trajectory in a radius of 25 km when monitored with a tolerance of one day backwards around the actual parcel location. A pixel was considered to contain fire, if the relevant temperature channel of the sensor detected average temperatures above 70°C. The firepixel data was obtained from the satellites NOAA-12 for the whole study period. To improve the accuracy of our fire pixel count we additionally used data of the satellites GOES-8 (Geostationary Environmental Satellite) and MODIS 01D (Moderate Resolution Imaging Spectroradiometer) for the year 2002. A fire pixel was considered as identified, if it was classified as such by at least one of these three satellites. The fire pixel data are provided by the Brazilian Space Agency INPE (Instituto Nacional de Pesquisas Espaciais; http://paraguay.cptec.inpe.br/produto/queimadas/). More details of the fire pixel count are given in the work of [Rollenbeck R. et al., 2008]].

Afterwards we tested the consistency of the firepixel data with the GOME (Global Ozone Monitoring Experiment) NO_2 remote sensing product [Richter et al., 2005] for 4°S 79°W ± 2° (= the air column over the study site). Biomass burning periods were defined as the sum of all weeks when the GOME NO_2 index was higher than the 5-year mean and more than 14 firepixels (25%-percentile) were counted on the pathway of this week's trajectories (Figure C-1).

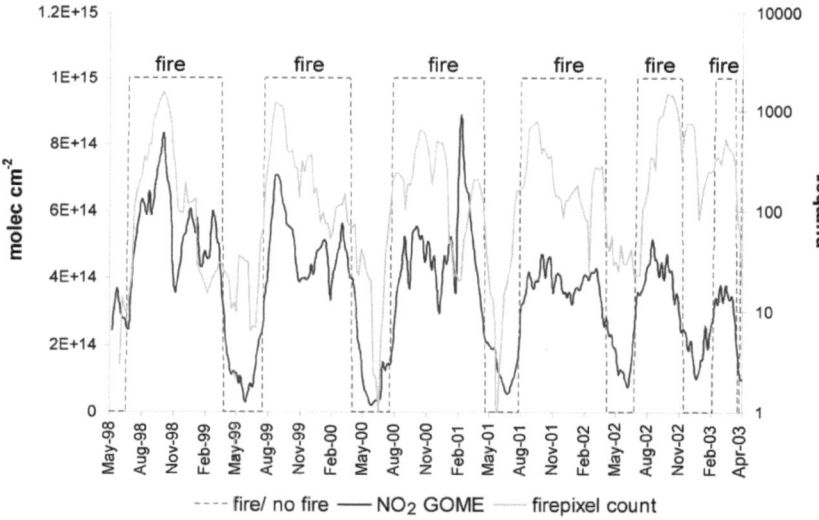

Figure C-1: Classification of biomass-burning periods using NO_2-index satellite data of GOME (Global Ozone Monitoring Experiment) in the air column above the study site (4°S 79°W ± 2°) and fire pixel count of NOAA 12 (National Oceanic and Atmospheric Administration) in Amazonia and northern South America along the daily Hysplit trajectories (radius 25 km around the actual parcel location).

4. Results and Discussion

4.1 Influence of Amazonian biomass burning on element deposition in the north Andes

In our test of the chemical stability of the precipitation samples, the weekly precipitation volume was 300 ml (s.e. 78 ml) in the daily emptied samplers and 322 ml (55) in the weekly samples. Weekly volume-weighted mean electrical conductivity and pH were 32 mS (11) and 6.5, respectively for the daily samples and 31 mS (8.4) and 6.7 for the weekly samples, respectively. The volume-weighted concentrations of NO_3-N, NH_4-N, and TOC were 0.06 (0.02), 0.72 (0.06), and 5.2 (1.1) mg l^{-1} in the daily emptied samplers and 0.12 (0.06), 0.74 (0.12), and 5.8 (1.0) in the weekly samples, respectively. There were no significant differences between the calculated mean weekly chemical composition of the daily collected samples and the measured chemical composition of the cumulative weekly sample (Tukey's HSD test, n = 6). Thus, there are no indications for post-sampling chemical transformations in our collectors.

The bulk deposition of most studied elements with rainfall was at the lower end of published ranges for similar ecosystems (Table C-1). N and Na deposition with rainfall was well in the middle of the range, while TOC, K, P, and Mn deposition was at the upper end of the range.

The temporal course of GOME NO_2 integrated column amounts paralleled those of the concentrations of total H^+, N, NO_3-N, Mn, and TOC in rainfall at our study site (Figure C-2). This coincides with the observation that N, Mn, and TOC deposition rates were at the upper end of published values in similar ecosystems (Table C-1). Ammonium is quickly oxidized in the atmosphere near the distant sources [*Asman et al.*, 1998] and therefore NH_3-N concentrations were not correlated with the GOME NO_2 column amounts. Interestingly, S did not show a direct relationship to biomass burning indicating the strong influence of local S sources at our study site like volcanism and biogenic aerosols. Both likely source-types of biomass burning-related aerosols downwind to our study site - tropical forest and savanna/grasslands - emit little SO_2 [*Andreae and Merlet*, 2001; *Yamasoe et al.*, 2000]. In the absence of long-range aerosol transport from the Sahara, a significant source of base metal deposition in 1999/2000 (framed in Figure C-2 [*Boy and Wilcke*, 2008]), K concentrations were also strongly influenced by biomass burning in the Amazon (Figure C-2. Our interpretations are supported by the results of a principal component analysis of element concentrations in

Table C-1: Comparison of annual bulk depositon (kg ha^{-1} yr^{-1}) with rainfall in selected tropical montane forests.

Bulk deposition	H$^+$	N$_{tot}$	NO$_3$-N	DON	NH$_4$-N	TOC	K	Ca	Mg	Na	P	S	Mn	Al	Zn	Reference
								kg ha^{-1} yr^{-1}								
Ecuador (montane forest)	0.12-0.16	9.5-10	2.6-2.9	4.7-5.1	2.5	102-117	7.2-8.3	5.6-5.7	2.2-2.4	14-16	0.64-1.1	2.3-2.9	0.030-0.037	0.12-0.16	0.22-0.27	This study
Central Amazon	0.30	-	3.1	-	1.7	-	1.6	2.0	-	2.2	-	11	-	-	-	Andreae et al. [1990] [Williams et al. 1997]
Central Amazon	0.47	-	7.1	-	1.5	-	0.9	2.6	0.61	1.5	0.22	5.2	-	-	-	Lilienfein and Wilcke [2004]
Cerrado	0.07-0.10	5.8-6.7	-	-	-	50-54	5.2-6.3	2.7-3.0	1.2-1.4	4.7-5.0	-	-	0.009-0.011	0.026-0.029	0.096-0.12	Da Rocha et al. [2005]
SE-Brazil	-	-	9.2	-	2.9	-	1.1	1.7	0.30	1.1	-	6.4	-	-	-	Mayer et al. [2000]
SE-Brazil	0.32-2.2	-	6.6-10	-	5.2-43	-	11-27	13-76	3.4-16	20-31	-	41-110	-	1.8-5.2	-	
Colombia (montane forest)	-	-	-	-	11-18	-	6.9-7.9	7.3-10	2.5-3.2	16-24	0.48-0.72	17-26	-	-	-	Veneklaas [1990]
Venezuela (Maracaibo)	-	5.2-11	1.7-4.2	-	3.5-7.2	-	-	-	-	-	-	11-13	-	-	-	Morales et al. [1998]
Costa Rica (La Selva lowland forest)	0.20	-	1.8-3.4	1.0-6.6	2.6-5.0	22-36	2.2-5.0	3.9-9.0	3.0-4.9	23-33	-	7.5-11	-	-	-	Eklund et al. [1997]
Puerto Rico (Pico del Este)	0.32	-	5.3	-	6.9	-	27	47	30	247	-	38	-	-	-	Asbury et al. [1994]
Congo (Dominica)	0.28	-	8.1	-	1.7	-	1.2	2.8	-	3.8	-	5.0	-	-	-	Lacaux et al. [1992]
Congo (Boyele)	0.81	-	24	-	5.6	-	2.5	3.7	-	3.4	-	16	-	-	-	Lacaux et al. [1992]
Cameroon (Zoelele)	0.24	-	8.6	-	3.9	-	4.0	3.6	0.58	1.9	-	8.4	-	-	-	Sigha-Nkamdjou et al. [2003]
Nigeria (secondary lowland forest)	-	-	19-45	-	0.61-6	-	3.0-6.0	16-19	2.4-4.7	11-13	0.54-0.70	-	0	-	1.2-1.5	Maogbalu, [2003]
Cameroon (Korup NP)	-	-	-	-	-	-	7.8	9.3	5.3	-	-	-	-	-	-	Chuyong et al. [2004]
Tansania (Kilimanjaro)	-	-	-	3.4-6.2	-	59-144	-	-	-	-	-	-	-	-	-	Schrumpf et al. [2006]
Cote d'Ivoire (Lamto wet Savanna)	0.08	-	5.7	-	3.8	-	1.1	2.3	0.39	1.7	-	3.7	-	-	-	Yoboue et al. [2005]
China (Ailao mountains)	-	10	0.91	-	2.7	-	3.0	8.0	3.2	1.7	-	2.8	-	-	-	Liu et al. [2003]

rainfall between May 1998 and April 2003 that grouped the analyzed elements into three classes: (i) the Saharan dust species Ca and Mg, (ii) the biomass-burning species TOC, total N, NO_3-N, Mn, and (iii) S, Na, and Cl$^-$ (Figure A-2). H^+ had similar loadings on PC1 than the biomass-burning species but loaded PC2 highly negatively. K, Al, and Zn were not assigned to any specific group probably because of their miscellaneous local and remote sources.

Significant differences in element concentrations of rainfall between "fire" and "no fire" periods existed for H^+, N_{tot}, NO_3-N, DON, TOC, and Mn. The concentrations of all other elements and compounds were not related to biomass burning periods (results not shown, overall ranges of bulk deposition rates are included in Table C-1). Element concentrations in rainfall were not or weakly correlated with rainfall volume, indicating that the variations in concentrations could not be explained by concentration/dilution effects (r = 0.001-0.175, n = 497-560). Differences in rainfall volume between the drier "fire" conditions (5 year-mean weekly rainfall of 34-41 mm for the three catchments, Table C-2) and the wetter "no fire" conditions (5 year-mean weekly rainfall of 62-63 mm) resulted in no significant differences in bulk deposition of the biomass burning-related elements between the "fire" and "no fire" periods. However, the drier conditions during "fire" periods also resulted in significantly higher dry deposition of biomass burning-related elements (Table C-3). Consequently, total depositions of NO_3-N, N_{tot}, DON, H^+, and Mn were higher during "fire" than "no fire" periods (Figure C-3). Total P deposition also tended to be higher during "fire" than "no fire" periods although differences were not significant. In a range of publications, the release of NO_3-N, N_{tot}, DON, H^+, Mn and P during biomass burning in Amazonia was reported [*Artaxo et al.*, 2002; *Hoffer et al.*, 2006; e.g., *Mace et al.*, 2003; *Mahowald et al.*, 2005; *Trebs et al.*, 2004; *Williams et al.*, 1997]. All these compounds showed higher total deposition at our study site during „fire" than during "no fire" conditions. Thus, biomass burning in Amazonia is a major driving factor of element deposition even to distant montane forest at the outer rim of Amazonia.

Biomass burning intensity is linked to the El Niño Southern Oscillation (ENSO) [e. g., *Alencar et al.*, 2006; *van der Werf et al.*, 2006]. Taking the proposed frequency shift of future ENSO due to global warming into account [*Timmermann et al.*, 1999] as well as growing interest in soybean cropping [*Soares et al.*, 2006; *Arima et al.*, 2007], our findings suggests rising deposition rates of biomass burning-related aerosols to our study site in the future.

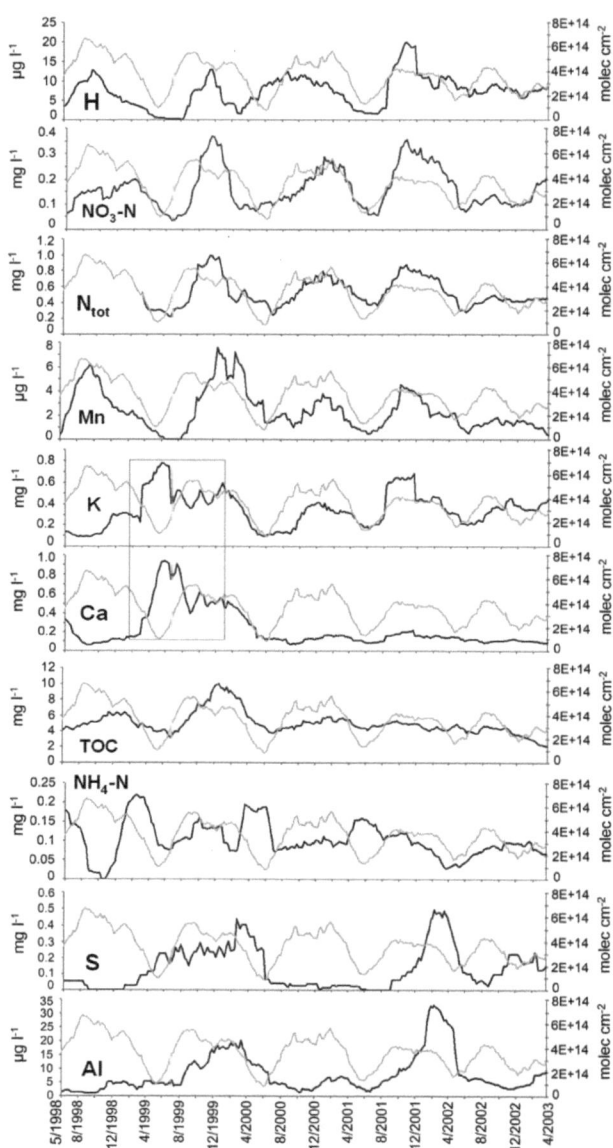

Figure C-2: Temporal courses of rainfall concentrations of selected chemical species (dark line, running 3-months average) in comparison to the NO_2-index of GOME (light line) as indication of fire acitivity in the Amazon basin between May 1998 and April 2003. The temporal course of Ca concentrations also represents those of Mg, which are closely correlated with the Ca concentrations (r=0.85). The box indicates a period of Saharan dust deposition in 1999/2000 which caused elevated base metal deposition.

Table C-2: Water fluxes of rainfall (RF), throughfall (TF), stemflow (ST), litter leachate (LL), soil solution at the 0.15 m and 0.3 m depths in mineral soil (SS15 + 30), and stream water (SW) of an Andean montane forest during biomass burning in Amazonia ("fire") and without ("no fire") between May 1998 and April 2003 for the five transects (T1- T3).

5- year mean water flux	unit	T1	T2.1	T2.2	T2.3	T3
		[mm week^{-1}]				
RF	fire	36	34	-	-	41
	no fire	62	62	-	-	63
TF	fire	18	22	21	20	22
	no fire	31	39	38	43	43
ST	fire	0.32	0.32	-	-	0.28
	no fire	0.66	0.67	-	-	0.65
LL	fire	13	18	16	15	17
	no fire	27	30	33	39	38
SS 15	fire	14	16	17	15	18
	no fire	28	26	33	39	41
SS 30	fire	13	17	16	15	16
	no fire	26	28	33	38	38
SF	fire	12	14	-	-	13
	no fire	24	32	-	-	29

4.2 Effect of deposited elements on the nutrient budgets of the Andean montane forest

Almost all deposited H$^+$ was buffered in the canopy (Figure C-4) both during "fire" and "no fire" conditions (Figure C-3). This effect is well known from acid rain and forest decline studies [e. g., *Matzner*, 2004; *Zeng et al.*, 2005]. Total deposition of base metals tended to be lower during "fire" conditions than during "no fire" conditions (Figure C-3), but base metal concentrations in throughfall were higher during "fire" than during "no fire" because of the enhanced H$^+$ buffering in the canopy during "fire". In MC 2, where the soils had the lowest base saturation (Figure C-5, a2), Ca and Mg were even retained in the canopy during "no fire" conditions (Figure C-4). This can be interpreted as indication of a growth limitation by Ca and Mg at particularly acid soils. This assumption is supported by the finding that Ca and Mg were immobilized during incubation of organic layer samples from MC2 [*Wilcke et al., 2002*]. In other studies in our working area, *Wilcke et al.* (2008) also observed a signficant effect of soil Ca concentrations on tree growth while *Soethe* et al. (2008) did not detect insufficient nutrient supply with the help of leaf analyses at 1900 a.s.l. but increasingly lower concentrations of all macronutrients with increasing altitude up to 3000 m (far above the upper limit of our study catchments).

Table C-3: Dry deposition at an Andean montane forest during biomass burning in Amazonia ("fire") and without ("no fire") between May 1998 and April 2003 for the five transects (T1-T3). Different lower case letters indicate significant differences of mean concentrations between "fire" and "no fire" (Games Howell p<0.05).

Dry deposition		H^+	N_{tot}	NO_3-N	DON	NH_4-N	TOC	K	Ca	Mg	Na	P	S	Mn	Al	Zn
	unit									g ha^{-1} week^{-1}						
T1	fire	3.4a	288a	100a	142a	58a	2320a	206a	94a	42a	444a	27a	56a	1.0a	3.6a	7.1a
	no fire	1.7b	124b	23b	61b	43a	1450b	120a	102a	43a	182b	8.6a	42a	0.28b	1.5b	2.8b
T2.1	fire	5.6a	333a	121a	169a	67a	3190a	188a	106a	39a	494a	24a	57a	1.4a	5.9a	7.6a
	no fire	1.3b	144b	27b	70b	51a	1810b	137a	133a	58a	242b	11a	53a	0.45a	2.4a	3.6b
T2.2	fire	3.4a	202a	73a	102a	41a	1930a	114a	64a	24a	299a	14a	34a	0.83a	3.6a	4.6a
	no fire	0.39b	44b	8.2b	21b	16b	550b	42b	40a	18a	74b	3.2a	16a	0.14b	0.73b	1.1b
T2.3	fire	4.4a	261a	95a	132a	53a	2490a	147a	83a	31a	386a	19a	45a	1.1a	4.6a	5.9a
	no fire	1.2b	128a	24b	62a	46a	1610a	123a	118b	52b	216a	9.5a	47a	0.40a	2.1a	3.2a
T3	fire	7.3a	372a	127a	186a	67a	3490a	252a	113a	42a	556a	55a	70a	1.5a	5.1a	11a
	no fire	2.0b	161b	33b	78b	55a	2140b	139a	135a	55a	266b	18a	76a	0.47b	2.3b	4.3b

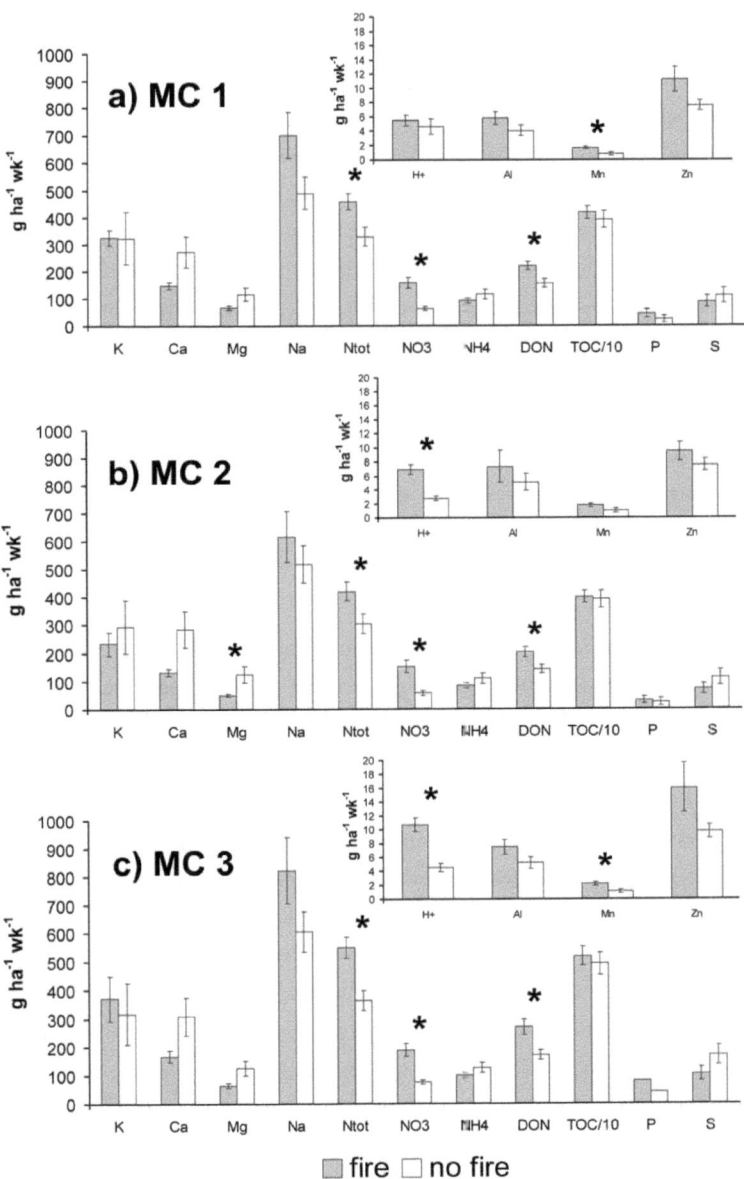

Figure C-3: Total deposition during biomass burning periods ("fire") compared to normal conditions ("no fire") between May 1998 and April 2003 for the microcatchments (MC) 1-3. Error bars represent standard errors and stars indicate significance (<0.05, Games Howell).

Elevated N deposition during "fire" periods increased concentrations in throughfall (Table C-3 and Figure C-2). Nevertheless, N was retained in the canopy during "fire" conditions (Figure C-4) but not during "no fire" conditions, when N deposition was much smaller (and therefore should even be more strongly retained if N was the only limiting nutrient). We offer two possible explanations for this observation. The first explanation is that forest growth at our study site is limited by several nutrients including N simultaneously ("co-limitation") and only the arrival of all these nutrients, which seems to be the case during "fire" conditions, result in a growth effect associated with increased nutrient uptake. This explanation is supported by findings of *Soethe et al.* [2008] based on nutrient analyses in leaves at a study site at 2400 m a.s.l. shortly above the upper limit of our study catchments. The second explanation is that N is not limiting but is taken up, if the limiting element is deposited at an increased rate because of the generally higher nutrient demand. There are some indications that Mn might be this limiting nutrient because (i) Mn deposition is higher during "fire" than "no-fire" conditions, (ii) Mn is consistently retained in the canopy (Figure C-4) and (iii) both the total deposition of Mn and the net Mn canopy budget correlate closely with the N_{tot} canopy budget (Figure C-6).

Canopy budgets of S were sensitive to biomass burning (Figure C-4), although S deposition did not differ significantly between the "fire" and "no fire" conditions (Figure C-3). In all microcatchments, S was leached from the canopy during "fire" conditions and retained during "no fire" conditions, suggesting that S and base metal response to fire-affected deposition was coupled in the canopy. Possibly, sulfate accompanied the cationic base metals because of charge compensation.

Elevated base metal leaching during "fire" conditions was also observed in the litter leachates below the organic horizon and to a lesser degree at 0.15 m depth in the mineral soil (Table C-4, Figure C-5a). The differences in total base metal loss from the 0.15 m depth in the mineral soil among the transects (6.7-44 mol_c ha^{-1} wk^{-1}) can be attributed to the differences in base saturation among the transects (6.3-95%, Figure C-5a1-a5). Since 67% and 15% of the nutrient-absorbing roots are located in the O horizon and upper 0.15 m of the mineral soil, respectively [*Soethe et al.*, 2006], this loss is a nutrient depletion for the vegetation (in spite of retention in the subsoil).

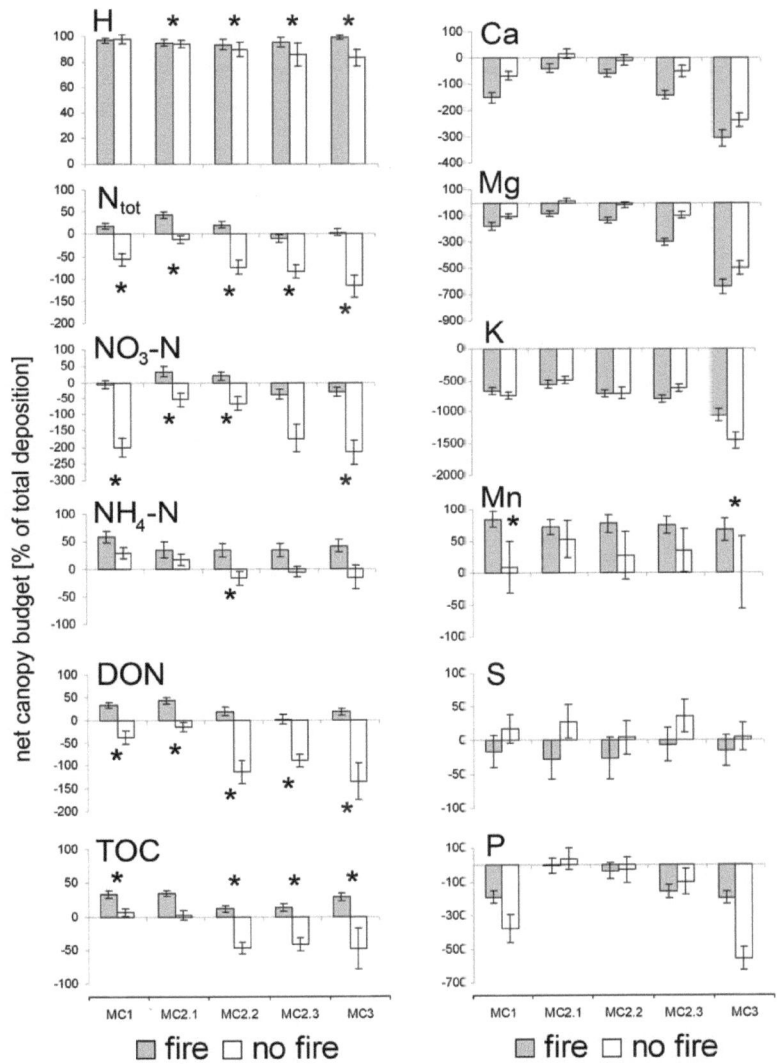

Figure C-4: Net canopy budget expressed in percent of total deposition between May 1998 and April 2003 for the microcatchments (MC) 1-3. Negative values indicate net loss, positive net retention of the canopy. Error bars represent standard errors. Stars indicate significant differences between "fire" and "no fire" canopy budgets (<0.05, Games Howell; statistical test based on fluxes, not percentage of total deposition).

At the 0.3 m depth and in stream water, there were no significant differences between "fire" and "no fire" conditions because of the large retention capacity of the mineral soil (Figure C-5a). Thus, the fire-derived acidification front reached, on average, the 0.15 m depth of the mineral soil between 1998 and 2003.

NO_3-N and N_{tot} fluxes in the O horizon also indicate N loss for all of the catchments (Figure C-5b, c). N losses were much higher during "no fire" conditions than during "fire" conditions, thus indicating (as in the canopy) that N is only retained during periods with generally elevated nutrient deposition.

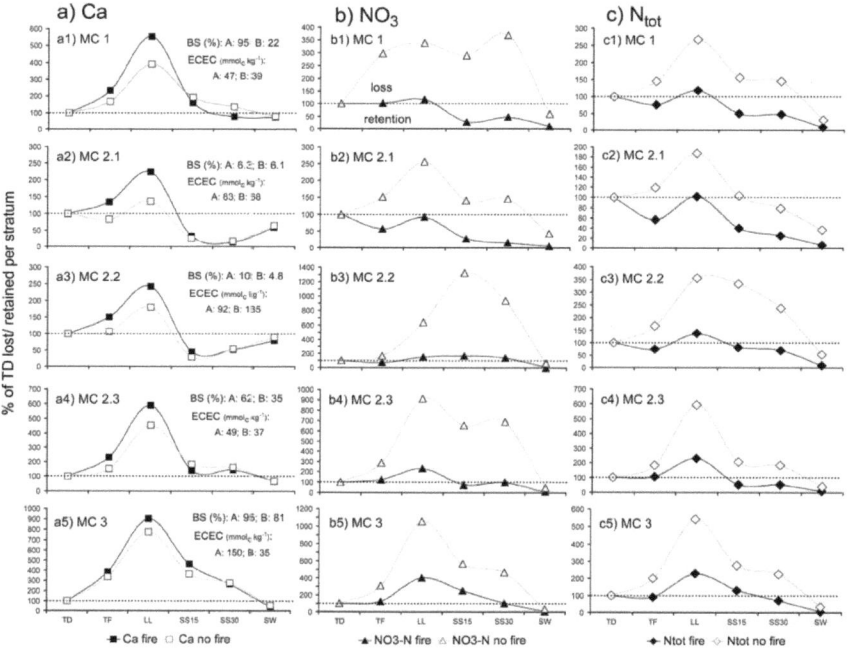

Figure C-5: Percentage of total deposition being lost/retained down to a given stratum [LL= flux in litter leachate; SS15 and SS30= fluxes at the 0.15 and 0.3 m depths in mineral soil, respectively; SW= flux in surface flow (stream water)] between May 1998 and April 2003 for the microcatchments (MC) 1-3. For each transect base saturation (BS) and cation exchange capacity (ECEC) is given in column a). Direction of loss or retention is indicated in pannel b1).

Table C-4: Volume-weighted mean concentrations in litter leachate (LL), soil solution at the 0.15 m and 0.3 m depths in mineral soil (SS15 + 30), and stream water (SW) of an Andean montane forest during biomass burning in Amazonia ("fire") and without ("no fire") between May 1998 and April 2003 for the five transects (T1- T3). Different lower case letters indicate significant differences of mean concentrations between "fire" and "no fire" (Games Howell p<0.05).

Concentrations			H^+	Ntot	NO_3-N	DON	TOC	K	Ca	Mg	Na	Mn
		units	[ug l^{-1}]	---[mg l^{-1}]---								[ug l^{-1}]
LL	T1	fire	3.7a	4.1a	1.4a	2.4a	48a	8.9a	6.3a	2.0a	1.7a	12a
		no fire	2.5a	3.3a	0.79a	2.1a	31b	5.2a	4.0a	1.3b	1.5a	7.5a
	T2.1	fire	28a	2.7a	0.92a	1.4a	40a	5.3a	1.9a	1.2a	1.3a	47a
		no fire	33a	2.1a	0.52a	1.2a	35b	2.6b	1.4a	1.0a	0.74a	29b
	T2.2	fire	6.0a	3.0a	1.2a	1.5a	34a	8.0a	1.7a	1.1a	1.2a	21a
		no fire	7.6a	2.6a	0.87a	1.3a	25b	7.5a	1.2a	0.69a	0.82a	8.0a
	T2.3	fire	3.3a	6.2a	2.2a	3.3a	60a	16a	5.0a	2.2a	2.7a	21a
		no fire	5.7a	5.0a	1.4a	2.7a	45b	12a	3.5b	1.5b	1.9a	20a
	T3	fire	0.8a	7.4a	4.5a	2.4a	58a	18a	8.9a	5.4a	1.6a	8.0a
		no fire	3.4a	5.1b	2.0b	2.5a	44b	11a	6.2b	3.6b	1.0a	5.5a
SS 15	T1	fire	9.5a	1.6a	0.32a	1.2a	20a	1.3a	1.6a	0.58a	0.70a	16a
		no fire	4.0a	1.8a	0.63a	1.1a	18a	1.4a	1.8a	0.63a	0.33b	15a
	T2.1	fire	70a	1.3a	0.33a	0.82a	26a	0.64a	0.33a	0.19a	0.69a	5.7a
		no fire	66a	1.3a	0.32a	0.92a	29b	0.74a	0.30b	0.16b	0.37a	5.3a
	T2.2	fire	36a	1.7a	1.3a	0.39a	7.9a	0.81a	0.29a	0.56a	0.68a	64a
		no fire	37a	2.4a	1.8a	0.55a	7.2a	0.74a	0.20a	0.64a	0.23b	66a
	T2.3	fire	14a	1.4a	0.75a	0.55a	11a	0.46a	1.2a	0.37a	0.56a	13a
		no fire	14a	1.7a	1.01a	0.58b	11a	0.32a	1.4b	0.45b	0.31a	18a
	T3	fire	4.0a	3.9a	2.6a	1.2a	19a	1.8a	4.2a	2.7a	1.6a	11a
		no fire	0.68a	2.5b	1.0a	1.3a	19a	1.6a	2.8b	2.0b	0.53a	11a
SS 30	T1	fire	5.9a	1.7a	0.58a	0.99a	15a	1.0a	0.90a	0.36a	1.3a	7.5a
		no fire	7.7a	1.8a	0.88a	0.84a	14b	1.1a	1.4b	0.58b	0.38b	11a
	T2.1	fire	40a	0.70a	0.16a	0.49a	15a	0.16a	0.16a	0.09a	0.63a	8.0a
		no fire	44a	0.93a	0.29a	0.56a	16a	0.20a	0.19a	0.13a	0.25a	7.6a
	T2.2	fire	29a	1.5a	1.1a	0.35a	7.7a	0.88a	0.36a	0.59a	1.05a	44a
		no fire	29a	1.7a	1.3a	0.37a	7.6a	1.1a	0.37a	0.74a	0.29b	60a
	T2.3	fire	7.7a	1.4a	0.95a	0.39a	7.8a	0.56a	1.2a	0.36a	0.76a	13a
		no fire	6.3a	1.6a	1.1a	0.38a	7.0a	0.41a	1.3a	0.49a	0.29a	14a
	T3	fire	0.9a	2.3a	1.2a	1.0a	14a	1.2a	2.7a	2.1a	1.1a	5.2a
		no fire	0.56b	2.2b	0.90a	1.1b	15b	0.59a	2.3a	2.1a	0.68a	8.7a
SW	T1	fire	0.26a	0.32a	0.13a	0.10a	3.1a	0.64a	0.92a	0.38a	3.9a	0.65a
		no fire	0.26a	0.43a	0.15a	0.12a	4.7b	0.44a	0.89a	0.41a	3.2a	0.55a
	T2	fire	0.17a	0.22a	0.07a	0.07a	3.3a	0.32a	0.62a	0.40a	3.2a	0.72a
		no fire	0.26a	0.40b	0.09a	0.13a	5.7b	0.37a	0.64a	0.47a	2.5b	0.76a
	T3	fire	0.27a	0.28a	0.09a	0.11a	4.5a	0.84a	0.52a	0.38a	2.5a	2.0a
		no fire	3.2b	0.43b	0.08a	0.18b	8.4b	0.51a	0.59a	0.33a	1.8b	1.8a

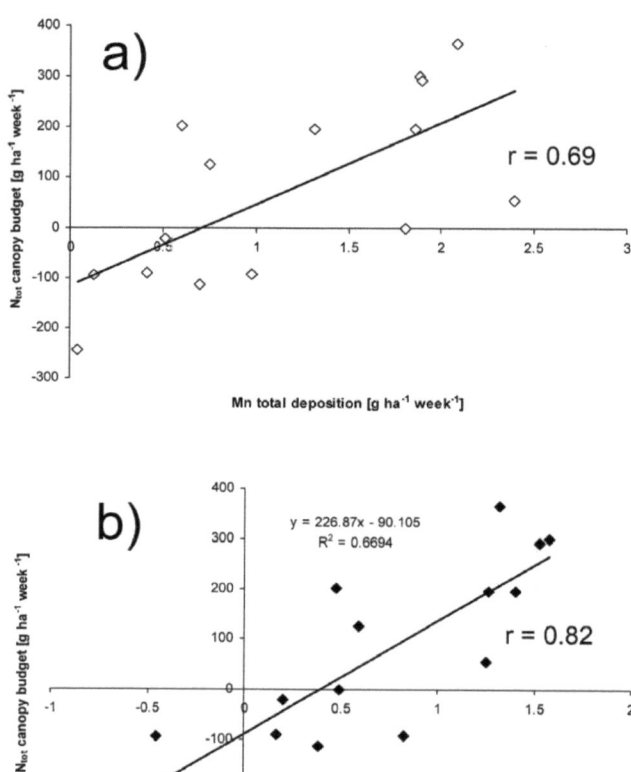

Figure C-6: Correlations between Mn total deposition and N_{tot}-canopy budget (a), and between Mn- and N_{tot} canopy budget (b) for May 1999-April 2003 (1998 no N_{tot} measured) in transect MC 2.1.

5. Conclusions

We observed a substantially elevated deposition of H^+, NO_3, N_{tot}, DON, Mn, and TOC during the biomass burning season in Amazonia and upwind which was at the upper end of published values for similar ecosystems except for H^+.

Elevated H^+ deposition caused a loss of base metals from the canopy, the organic layer, and the mineral toposil, deteriorating base metal supply of the vegetation. Thus, there is a clear acidification of our study ecosystem caused by Amazonian forest fires. The significant net retention of N and Mn in the canopy and soil suggests a fertilizing effect of the forest fires possibly resulting in eutrophication of the studied nutrient-poor forest ecosystem. However, the nutritional status of the studied tropical montane forest is still unclear and can only be solved with nutrient addition experiments.

Both acidification and eutrophication likely have an impact on the floristic composition of the montane forests. As these forests belong to the most species-rich ecosystems of the world there is the risk of biodiversity loss.

We conclude that the current biomass burning activity in Latin America already is strong enough to interfere with nutrient cycling of the remotest parts of Amazonia. Thus, any changes in the intensity of biomass burning in Amazonia will have an effect on all Andean montane forests within the trade wind zone which receive element inputs from air masses passing the Amazon basin.

6. Acknowledgements

We are particularly grateful to Wolfgang Zech who initiated this long-term research project. We thank Paul Emck for providing meteorological data, Syafrimen Yasin for the soil data, K. Fleischbein, R. Goller, M. Sequeira and numerous students for help with data acquisition, A. Richter for providing the GOME data and valuable discussion, and the NOAA for admission of HYSPLIT data. This work was funded by the German Research Foundation (DFG FOR 402, Wi1601/5-1,-2,-3).

7. References

Alencar, A., D. Nepstad, and M. D. V. Diaz (2006), Forest understory fire in the Brazilian Amazon in ENSO and non-ENSO years: Area burned and committed carbon emissions, *Earth Interactions*, *10*, doi:10.1175/EI1150.1171.

Allen, A. G., and A. H. Miguel (1995), Biomass Burning in the Amazon - Characterization of the Ionic Component of Aerosols Generated Tom Flaming and Smoldering Rain-Forest and Savanna, *Environmental Science & Technology*, 29(2), 486-493.

Andreae, M. O., et al. (1988), Biomass-burning emissions and associated haze layers over Amazonia, *Journal of Geophysical Research* 93(D2), 1509-1527.

Andreae, M. O., and P. Merlet (2001), Emission of trace gases and aerosols from biomass burning, *Global Biogeochemical Cycles*, 15(4), 955-966.

Artaxo, P., J. V. Martins, M. A. Yamasoe, A. S. Procopio, T. M. Pauliquevis, M. O. Andreae, P. Guyon, L. V. Gatti, and A. M. C. Leal (2002), Physical and chemical properties of aerosols in the wet and dry seasons in Rondonia, Amazonia, *Journal of Geophysical Research-Atmospheres*, 107(D20), 8081, doi:8010.1029/2001JD000666.

Asman, W. A. H., M. A. Sutton, and J. K. Schjorring (1998), Ammonia: emission, atmospheric transport and deposition, *New Phytologist*, 139(1), 27-48.

Bendix, J., R. Rollenbeck, D. Gottlicher, and J. Cermak (2006), Cloud occurrence and cloud properties in Ecuador, *Climate Research*, 30(2), 133-147.

Beven, K. J., R. Lamb, P. F. Quinn, and R. B. Romanowicz (1995), TOPMODEL, in *Computer Models of Watershed Hydrology*, edited by V. P. Singh, pp. 627-668, Water Resource Publications, Colorado.

Boy, J., and W. Wilcke (2008), Tropical Andean forest derives calcium and magnesium from Saharan dust, *Global Biogeochemical Cycles*, 22(1), doi:10.1029/2007GB002960.

Brooks, T. M., et al. (2002), Habitat loss and extinction in the hotspots of biodiversity, *Conservation Biology*, 16(4), 909-923.

Chadwick, O. A., L. A. Derry, P. M. Vitousek, B. J. Huebert, and L. O. Hedin (1999), Changing sources of nutrients during four million years of ecosystem development, *Nature*, 397(6719), 491-497.

Clark, K. L., N. M. Nadkarni, D. Schaefer, and H. L. Gholz (1998), Atmospheric deposition and net retention of ions by the canopy in a tropical montane forest, Monteverde, Costa Rica, *Journal of Tropical Ecology*, 14, 27-45.

Da Rocha, G. O., A. G. Allen, and A. A. Cardoso (2005), Influence of agricultural biomass burning on aerosol size distribution and dry deposition in southeastern Brazil, *Environmental Science & Technology*, 39(14), 5293-5301.

Dixon, R. K., and J. Wisniewski (1995), Global forest systems: An uncertain response to atmospheric pollutants and global climate change?, *Water Air and Soil Pollution*, 85(1), 101-110.

Draxler, R. R., and G. D. Hess (1998), An overview of the HYSPLIT_4 modelling system for trajectories, dispersion and deposition, *Australian Meteorological Magazine*, 47(4), 295-308.

DVWK (Deutscher Verband für Wasserwirtschaft und Kulturbau) (1996), *Ermittlung der Verdunstung von Land- und Wasserflächen*, DVWK, Bonn.

Fabian, P., M. Kohlpaintner, and R. Rollenbeck (2005), Biomass burning in the Amazon-fertilizer for the mountaineous rain forest in Ecuador, *Environmental Science and Pollution Research*, 12(5), 290-296.

Fleischbein, K., W. Wilcke, R. Goller, J. Boy, C. Valarezo, W. Zech, and K. Knoblich (2005), Rainfall interception in a lower montane forest in Ecuador: effects of canopy properties, *Hydrological Processes*, 19(7), 1355-1371.

Fleischbein, K., W. Wilcke, C. Valarezo, W. Zech, and K. Knoblich (2006), Water budgets of three small catchments under montane forest in Ecuador: experimental and modelling approach, *Hydrological Processes*, 20(12), 2491-2507.

Freitas, S. R., K. M. Longo, M. A. F. S. Diasb, P. L. S. Diasb, R. Chatfield, E. Prins, P. Artaxo, G. A. Grell, and F. S. Recuero (2005), Monitoring the transport of biomass burning emissions in South America, *Environmental Fluid Mechanics*, 5(1-2), 135-167.

Generoso, S., F. M. Breon, Y. Balkanski, O. Boucher, and M. Schulz (2003), Improving the seasonal cycle and interannual variations of biomass burning aeroscl sources, *Atmospheric Chemistry and Physics*, 3, 1211-1222.

Hoffer, A., A. Gelencser, M. Blazso, P. Guyon, P. Artaxo, and M. O. Andreae (2006), Diel and seasonal variations in the chemical composition of biomass burning aerosol, *Atmospheric Chemistry and Physics*, 5, 3505-3515.

Ichoku, C., and Y. J. Kaufman (2005), A method to derive smoke emission rates from MODIS fire radiative energy measurements, *Ieee Transactions on Geoscience and Remote Sensing*, 43(11), 2636-2649.

Jaegle, L., L. Steinberger, R. V. Martin, and K. Chance (2005), Global partitioning of NOx sources using satellite observations: Relative roles of fossil fuel combustion, biomass burning and soil emissions, *Faraday Discussions*, 130, 407-423.

Kauffman, J. B., D. L. Cummings, D. E. Ward, and R. Babbitt (1995), Fire in the Brazilian Amazon .1. Biomass, Nutrient Pools, and Losses in Slashed Primary Forests, *Oecologia*, 104(4), 397-408.

Likens, G. E., and J. S. Eaton (1970), A polyurethane stemflow collector for trees and shrubs, *Ecology*, 51, 938-939.

Lloyd, C. R., and A. Marques (1988), Spatial variability of throughfall and stemflow measurements in Amazonian rain forest, *Agricultural and Forest Meteorology*, 47, 63-73.

Mace, K. A., P. Artaxo, and R. A. Duce (2003), Water-soluble organic nitrogen in Amazon Basin aerosols during the dry (biomass burning) and wet seasons, *Journal of Geophysical Research-Atmospheres*, 108(D16), 4512, doi:4510.1029/2003JD003557.

Mahowald, N. M., P. Artaxo, A. R. Baker, T. D. Jickells, G. S. Okin, J. T. Randerson, and A. R. Townsend (2005), Impacts of biomass burning emissions and land use change on Amazonian atmospheric phosphorus cycling and deposition, *Global Biogeochemical Cycles*, 19(4), GB4030, doi:4010,1029/2005GB002541.

Martinelli, L. A., M. C. Piccolo, A. R. Townsend, P. M. Vitousek, E. Cuevas, W. McDowell, G. P. Robertson, O. C. Santos, and K. Treseder (1999), Nitrogen stable isotopic composition of leaves and soil: Tropical versus temperate forests, *Biogeochemistry*, 46(1-3), 45-65.

Matzner, E. (2004), *Biogeochemistry of Forested Catchments in a changing Environment. A German case study*, Springer, Berlin.

Parker, G. G. (1983), Throughfall and stemflow in the forest nutrition cycle, *Advances in Ecological Research*, 13, 57-133.

Pereira, E. B., A. W. Setzer, F. Gerab, P. E. Artaxo, M. C. Pereira, and C. Monroe (1996), Airborne measurements of aerosols from burning biomass in Brazil related to the TRACE a experiment, *Journal of Geophysical Research-Atmospheres*, 101(D19), 23983-23992.

Phoenix, G. K., et al. (2006), Atmospheric nitrogen deposition in world biodiversity hotspots: the need for a greater global perspective in assessing N deposition impacts, *Global Change Biology*, 12(3), 470-476.

Richter, A., J. P. Burrows, H. Nuss, C. Granier, and U. Niemeier (2005), Increase in tropospheric nitrogen dioxide over China observed from space, *Nature*, 437(7055), 129-132.

Rodhe, H., F. Dentener, and M. Schulz (2002), The global distribution of acidifying wet deposition, *Environmental Science & Technology*, *36*(20), 4382-4388.

Rollenbeck R., Fabian P., and Bendix J. (2008), Temporal Heterogeneities - Matter deposition from remote areas, in *Gradients in a tropical mountain ecosystem of Ecuador*, edited, pp. 303-310.

Soethe, N., J. Lehmann, and C. Engels (2006), The vertical pattern of rooting and nutrient uptake at different altitudes of a south ecuadorian montane forest, *Plant and Soil*, *286*(1-2), 287-299.

Soethe, N., W. Wilcke, J. Homeier, J. Lehmann, and C. Engels (2008), Plant growth along the altitudinal gradient - role of plant nutritional status, fine root activity, and soil properties, in *Gradients in a Tropical Mountain Ecosystem of Ecuador*, edited by E. Beck, et al., Springer, Heidelberg.

Timmermann, A., J. Oberhuber, A. Bacher, M. Esch, M. Latif, and E. Roeckner (1999), Increased El Nino frequency in a climate model forced by future greenhouse warming, *Nature*, *398*(6729), 694-697.

Trebs, I., F. X. Meixner, J. Slanina, R. Otjes, P. Jongejan, and M. O. Andreae (2004), Real-time measurements of ammonia, acidic trace gases and water-soluble inorganic aerosol species at a rural site in the Amazon Basin, *Atmospheric Chemistry and Physics*, *4*, 967-987.

Tsigaridis, K., and M. Kanakidou (2003), Global modelling of secondary organic aerosol in the troposphere: a sensitivity analysis, *Atmospheric Chemistry and Physics*, *3*, 1849-1869.

Ulrich, B. (1983), *Effects of Accumulation of Air Pollutants in Forest Ecosystems*, Reidel, Dordrecht.

van der Werf, G. R., J. T. Randerson, L. Giglio, G. J. Collatz, P. S. Kasibhatla, and A. F. Arellano (2006), Interannual variability in global biomass burning emissions from 1997 to 2004, *Atmospheric Chemistry and Physics*, *6*, 3423-3441.

Wardle, D. A., L. R. Walker, and R. D. Bardgett (2004), Ecosystem properties and forest decline in contrasting long-term chronosequences, *Science*, *305*(5683), 509-513.

Whitehead, H. L., and J. H. Feth (1964), Chemical composition of rain, dry fallout and bulk precipitation at Menlo Park, California, *Journal of Geophysical Research*, *69*, 3319-3333.

Williams, M. R., T. R. Fisher, and J. M. Melack (1997), Chemical composition and deposition of rain in the central Amazon, Brazil, *Atmospheric Environment*, *31*(2), 207-217.

Yamasoe, M. A., P. Artaxo, A. H. Miguel, and A. G. Allen (2000), Chemical composition of aerosol particles from direct emissions of vegetation fires in the Amazon Basin: water-soluble species and trace elements, *Atmospheric Environment*, *34*(10), 1641-1653.

Yasin, S. (2001), Water and nutrient dynamics in microcatchments under montane rain forest in the South Ecuadorian Andes, Ph. D. thesis, University of Bayreuth, Bayreuth, Germany.

Zeng, G. M., G. Zhang, G. H. Huang, Y. M. Jiang, and H. L. Liu (2005), Exchange of Ca^{2+}, Mg^{2+} and K^+ and uptake of H^+, NH_4^+ for the subtropical forest canopies influenced by acid rain in Shaoshan forest located in Central South China, *Plant Science*, *168*(1), 259-266.

D Water flow paths in soil control element exports in an Andean tropical montane forest[a]

Jens Boy[1], Carlos Valarezo[2] & Wolfgang Wilcke[1]

[1] Geographic Institute, Johannes Gutenberg University, 55099 Mainz, Germany
[2] Universidad Nacional de Loja, Area Agropecuaria y de Recursos Naturales Renovables, Programa de Agroforestería, Ciudadela Universitaria Guillermo Falconí, Loja, Ecuador

[a] European Journal of Soil Science 59, 1209-1227, 2008

1. Abstract

We tested the hypothesis that concentrations of chemical constituents in stream water can be explained by the depth of water flow through soil. Therefore, we measured the concentrations of total organic carbon (TOC), NO_3-N, NH_4-N, dissolved organic nitrogen (DON), P, S, K, Ca, Mg, Na, Al, and Mn in rainfall, throughfall, stemflow, litter leachate, mineral soil solution and stream water of three 8-13 ha catchments on steep slopes (1900-2200 m a.s.l.) of the south Ecuadorian Andes, from April 1998-April 2003. Peak C (14–22 mg litre^{-1}), N (0.6-0.9 mg litre^{-1}), K (0.5-0.7 mg litre^{-1}), Ca (0.6-1.0 mg litre^{-1}), Mg (0.3-0.5 mg litre^{-1}), Al (110-390 µg litre^{-1}) and Mn (3.9-8.4 µg litre^{-1}) concentrations in stream water were associated with lateral flow (fast near-surface flow in saturated topsoil) while the greatest P (0.1-0.3 mg litre^{-1}), S (0.3-0.7 mg litre^{-1}) and Na (3.0-6.0 mg litre^{-1}) concentrations occurred during low baseflow conditions. All elements had greater concentrations in the organic layer than in the mineral soil, but only C, N, K, Ca, Mg, Al and Mn were flushed out during lateral-flow conditions. Phosphorus, S and Na, in contrast, were mainly released by weathering and (re-)oxidation of sulfides in the subsoil. Baseflow accounted for 32% to 61% of P export, while >50% of S was exported during intermediate flow conditions, i.e. lateral flow at the depth of several tens of cm in the mineral soil. Near-surface water flow through C- and nutrient-rich topsoil during rainstorms was the major export pathway for C, N, Al, and Mn (contributing >50% to the total export of these elements). Near-surface flow also accounted for one third of total base metal export. Our results demonstrate that storm-event related near-surface flow markedly affects the cycling of many nutrients in steep tropical montane forests.

Abbreviations: TOC, total organic carbon; DON, dissolved organic nitrogen; DOC, dissolved organic carbon; TDN, total dissolved nitrogen; TDP, total dissolved phosphorus; TDS, total dissolved phosphorus; DOM, dissolved organic matter; MC, microcatchment; FDR, frequency domain reflectometry; AAS, atomic absorption spectroscopy; ECEC, effective cation exchange capacity; BS, base saturation; VWM, volume weighted mean; FWM, flow weighted mean; FE, flow class effect; ENSO, El Niño Southern Oscillation

2. Introducion

Nutrient budgets of forested catchments require characterization of hydrologically-mediated nutrient export. Determining nutrient export is complicated by the great variability of element concentrations in discharge waters [*Godsey et al.*, 2004; *McDowell and Asbury*, 1994].

In many studies, positive correlations between the discharge rate and element concentrations were observed, e.g. for dissolved organic carbon [DOC, *Buffam et al.*, 2001; *Hook and Yeakley*, 2005] or total dissolved nitrogen [TDN, *Campbell et al.*, 2000; *Goller et al.*, 2006]. Particularly during stormflow conditions, DOC, DON, and NO_3-N concentrations peaked and contributed up to 50% of total export in an Appalachian catchment [*Buffam et al.*, 2001]. In preceding studies at our site in southern Ecuador, increasing concentrations of DOC, DON, NO_3-N, and partly of NH_4-N in stream water were reported when discharge was high. The concentrations of these chemical constituents were greatest during periods of heavy rain [*Goller et al.*, 2006; *Wilcke et al.*, 2001].

Studies on the discharge behaviour of metals are, with the exception of Al, scarce, particularly in the tropics [*Elsenbeer et al.*, 1994; *Grimaldi et al.*, 2004; *Wilcke et al.*, 2001; *Yusop et al.*, 2006]. This is surprising given the importance of metals for tropical forest nutrition [*Campo et al.*, 2000; *Hedin et al.*, 2003]. In Amazonia, for instance, there is an indication that in many cases neither N nor P but base metals are the growth-limiting factors because base metals are often in poor supply in the strongly weathered soils [*Cuevas and Medina*, 1988; *Martinelli et al.*, 1999; *Rollenbeck et al.*, 2007; *Wardle et al.*, 2004]. Therefore understanding of pathways and processes of metal export from forested catchments is crucial for understanding Amazonian forest nutrition.

It is still undecided whether an increase in concentrations of chemical constituents in stream water during storm conditions is attributable to the existence of pre-event waters like e.g., displaced groundwater [summarized by *Genereux*, 2004] or to event water [*Brown et al.*, 1999; *Schellekens et al.*, 2004]. For steep, forested hillslopes, there is evidence that rapid, near-surface, lateral water flow often occurs [*Schellekens et al.*, 2004]. Increasing DOC and TDN concentrations in stream waters during storm events have been attributed to flushing of C- and N-rich topsoil into the stream [*Hagedorn et al.*, 2000; *Hook and Yeakley*, 2005]. Other studies have emphasized the role of soil moisture [*Western et al.*, 2004] and soil type [*Elsenbeer*, 2001] for spatially heterogeneous hydrological processes in soil, e.g. near-surface flow. Previous work at our study site by means of a $\delta^{18}O$ approach showed that a large portion of water flow in soil during rainstorms indeed occurred laterally, whereas during baseflow conditions most of the stream water originated from the deeper mineral soil [*Goller et al.*, 2005]. Associated with the finding that concentrations of DOC, NO_3-N, NH_4-N, DON, S, and P at our study site are greatest in the organic soil layers [*Wilcke et al.*, 2002], this suggested a relation of element export to the water flow regime and particularly to the depth of water flow in the soil.

Water flow in soil is generally considered to be a major control in the export of inorganic N [*Mitchell et al.*, 2001], DOC and DON [*Hagedorn et al.*, 2000], total P [*Goller et al.*, 2006] and S [*Kaiser et al.*, 2001]. Retention and transformation of nutrients, and hydrologic flow paths through the upper soil, which determine the contact time with soil, are major controls of element transport through soil [*Michalzik et al.*, 2001]. The role of organic forms of N, P, and S have become of greater interest in recent times because of the finding that DON seems to contribute more to the N cycle of unpolluted watersheds than studies from the anthropogenically influenced northern hemisphere had suggested [*Goller et al.*, 2006; *Oyarzun et al.*, 2004; *Perakis and Hedin*, 2002]. The concentrations of organic compounds in stream water are likely to be particularly influenced by lateral flow, because soluble organic matter accumulates in the organic layer and topsoil. This accumulation of organic matter also influences metal mobility in soil, as it is closely related to pH and the concentrations of ligands forming soluble metal complexes. Particularly important ligands are contained in the dissolved organic matter [DOC, *Rieuwerts*, 2007]. Al and Mn in particular are mobilized and transported as organo-complexes [*Lorieri and Elsenbeer*, 1997].

Studies combining hydrological and biogeochemical approaches to elucidate processes of stormflow-related element export are scarce, particularly in the Tropics [*Saunders et al.*, 2006]. Hinton et al. [1998] stated that most studies on hydrological nutrient export were not only leaving phenomena like stormflow untouched, but also were undertaken for relatively short periods, complicating their extrapolation to hydrological years. Especially in steep, forested catchments responding quickly to storm events (within minutes or hours), stream chemistry is much more influenced by changing environmental conditions than would be the case in slower reacting systems of lesser slope [*Schellekens et al.*, 2004]. This further complicates the extrapolation of single storm event data collected at high resolution to longer periods as is done when using the "storm chasing" approach [*Buffam et al.*, 2001] or flow separation approaches, which have proven suitable for catchments of lesser slope and/or greater size [e.g. *Evans and Davies*, 1998]. In contrast, studies sampling at fixed intervals over longer periods, as is done in long-term ecosystem studies [e.g. *Matzner*, 2004], usually calculate element export by multiplying flow-weighted mean concentrations with cumulative flow and face difficulties in assessing the influence of storm events on element export. This problem might be solved by defining a classification key based on discharge rate in order to estimate the contribution of different flow regimes to element export for long time series.

The objective of our study was to investigate the relationships among the distribution of chemical constituents in soil, flow regime, and element export with stream water in three small water catchments under tropical montane forest in southern Ecuador.

We tested the following four hypotheses: (i) The concentrations of chemical constituents in stream water of steep, forested catchments are related to discharge levels in a way that is specific for each chemical constituent, (ii) the depth of water flow in soil determines the concentrations of chemical constituents in stream water, (iii) discharge level classification is a suitable tool to estimate the contribution of different flow regimes to element export from steep, forested catchments for long time series and (iv) storm events have a significant influence on catchment nutrient export.

3. Material and Methods

3.1 Study site

The study area is located on the eastern slope of the "Cordillera Real", the eastern Andean cordillera in southern Ecuador, facing the Amazon basin between the cities of Loja and Zamora at 4° 00` S and 79° 05` W. We selected three 30-50° steep and 8-13 ha large microcatchments (MC1-3) under montane forest at an altitude of 1900-2200 m above sea level (a.s.l.) for our study (Figure A-1). We installed our equipment in each MC along transects, about 20 m long with an altitude range of 10 m, on the lower part of the slope at 1900-1910 m a.s.l. (transects MC1, MC2.1, and MC3). Moreover, we installed extra instrumentation at 1950-1960 (MC2.2) and 2000-2010 m a.s.l. (MC2.3). All transects were located below a closed forest canopy and aligned downhill. Three unforested sites near these microcatchments were used for rainfall gauging. Gauging site 2 has existed since April 1998, gauging sites 1 and 3 were built in May 2000. All catchments drain *via* small tributaries into the Rio San Francisco, which flows into the Amazon basin.

Within the monitored period (April 1998 and April 2003) annual precipitation ranged between 2340 and 2667 mm. Additional climate data were available from a meteorological station [*Richter*, 2003] between MC 2 and 3 (Figure A-1). June tended to be the wettest month with 302 mm of precipitation on average, in contrast to 78 mm in each of November and January, the driest months. The mean temperature at 1950 m a.s.l. was 15.5 °C. The coldest month was July, with a mean temperature of 14.5 °C, the warmest November with a mean temperature of 16.6 °C.

Recent soils have developed on postglacial landslides or possibly from periglacial cover beds [*Wilcke et al.*, 2001]. Soils are Humic Eutrudepts on transect MC1, Humic Dystrudepts on transects MC2.1, MC2.2, and MC2.3, and Oxyaquic Eutrudepts on transect MC3. All soils are shallow, loamy-skeletal with much mica. The organic layer consisted of Oi, Oe, and frequently also Oa horizons and had a thickness between 2 and 43 cm [mean of 16 cm; *Wilcke et al.*, 2002]. The thickness increased with increasing altitude to give Histosols (mainly Terric Haplosaprists) above *c.* 2100 m. Selected soil properties are summarized in Table A-1, the methods used are briefly described below. The underlying bedrock consists of interbedding of Palaeozoic phyllites, quartzites and metasandstones.

Microcatchments 2 and 3 are entirely forested, whereas the upper part of microcatchment 1 has been used for agriculture until about ten years ago. This part is currently undergoing natural succession and is covered by grass and shrubs. The study forest can be classified as "Lower Montane Forest [*Bruijnzeel and Hamilton*, 2000]. More information on the composition of the forest can be found in Homeier [2004].

3.2 Field sampling

Water samples were collected between April 1998 and April 2003. Each gauging station for incident precipitation consisted of five samplers. Solution sampled by rainfall collectors was "bulk precipitation", since collectors were open to dry deposition between rainfall events [*Parker*, 1983]. However, the contribution of dry deposition to rainfall collectors was assumed to be small because of the small sampling area compared to the "aerosol trapping capacity" of the entire forest [*Parker*, 1983].

Each of the five transects was equipped with five throughfall collectors (in May 2000 three more collectors were added to each transect). All throughfall samplers had a fixed position that was arbitrarily chosen and evenly distributed along the transects. To move samplers after each sample collection, as suggested by Lloyd & Marques [1988], to improve the representativity of the sample would have resulted in unacceptable damage to the study forest that was only accessible on very steep machete-cleared and rope-secured paths. More information on our throughfall measurements can be found in Fleischbein *et al*. [2005].

In the lowermost transects of each catchment (MC1, MC2.1, MC3), five trees were equipped with stemflow collectors. The throughfall and stemflow samples were combined to one sample per transect in the field. We furthermore installed three collectors for litter leachate (water vertically percolating through the organic layer) at lower, central, and upper positions along the transects and three suction lysimeters for soil solution sampling at each of the 0.15 m and 0.30 m depths in the mineral soil at a selected position along the transect. A combined soil water sample for each transect was produced by bulking the individual samples directly in the field. Soil solution was sampled after May 2000 after equilibration of the lysimeters in the soil for four months. Stream water samples were taken weekly from the centre of the streams at the outlet of each catchment.

Throughfall and rainfall collectors consisted of fixed 1-litre polyethylene sampling bottles and circular funnels with a diameter of 115 mm. The opening of the funnel was 0.3 m above the soil. The collectors were equipped with table tennis balls to reduce evaporation. Incident rainfall collectors were additionally wrapped with aluminum foil to reduce the impact of radiation. Stemflow collectors were made of polyurethane foam and connected with plastic tubes to a 10-litre container [*Likens and Eaton*, 1970]. In each catchment, four trees of the uppermost canopy layer and one tree fern belonging to the second tree layer were used for stemflow measurements. The species were selected to be representative of the study forest although this was difficult because of its great plant diversity. A list of the selected species is given in Fleischbein *et al*. [2005]. Litter leachate was sampled by zero tension lysimeters, consisting of plastic boxes (0.20 x 0.14 m sampling area) covered with a polyethylene net (0.5 mm mesh width). The boxes were connected to 1-litre polyethylene sampling bottles with a plastic tube. The lysimeters were installed from a soil pit below the organic layer and parallel to the surface. The organic layer was not disturbed; most roots in the organic layer remained intact [*Wilcke et al.*, 2001]. Mineral soil solution was sampled by suction lysimeters (mullite suction cups, 1 µm ± 0.1 µm pore size) with a vacuum pump. The vacuum was held permanently and and adjusted to the matric potential. The lysimeters did not collect the soil solution quantitatively.

We sampled Oi, Oe, Oa, A, and B horizons of 47 soils, and the sites were distributed so as to represent the soils proportionally all three catchments (Wilcke *et al.*, 2002). Samples were taken from the walls of a soil pit in a manner to be representative for the horizons. Soil samples were air-dried. O horizon samples were additionally finely ground with a ball mill, mineral soil samples were sieved to <2mm. All samples were stored in closed plastic bags at room temperature until analysis.

3.3 Hydrological measurements

Rainfall, throughfall and stemflow were measured weekly by recording single volumes for each collector. Additionally, each catchment was equipped with a tipping bucket rain gauge (NovaLynx 260-2500, NovaLynx Corporation, Grass Valley, California, U.S.A.) to obtain better resolution data of throughfall volume. Due to frequent logger breakdowns and funnel blocking the data set was incomplete. Missing data were substituted by regression of precipitation data of the meteorological station on throughfall for each catchment (equations (1)-(3)).

$Throughfall_{MC1} = 0.651\ Rainfall_{metstation} + 2.483\ (R^2=0.83,\ n=263;\ 16/4/1998-14/5/2003)$

(1)

$Throughfall_{MC2} = 0.707\ Rainfall_{metstation} + 2.583\ (R^2=0.88,\ n=263;\ 16/4/1998-14/5/2003)$

(2)

$Throughfall_{MC3} = 0.831\ Rainfall_{metstation} + 0.2913\ (R^2=0.86,\ n=263;\ 16/4/1998-14/5/2003)$

(3)

To quantify surface flow, in April 1998 Thompson (V-notch) weirs (90°) with sediment basins were installed in the lower part of each catchment and water levels were recorded hourly with a pressure gauge (water-level sensor). Additionally, water levels were measured manually after the stream water samples were collected. The empirical equations (4)-(6) were used to convert water levels to surface flow.

$$q(MC1) = 0.0140 \, h^{2.5156} \quad (4)$$
$$q(MC2) = 0.0146 \, h^{2.5575} \quad (5)$$
$$q(MC3) = 0.0081 \, h^{2.7067} \quad (6)$$

where q is surface flow (litre second^{-1}] and h the water level [cm]. Equations (4)-(6) were derived from direct measurement of the surface flow at different water levels (MC1: n=31; MC2: n = 28; MC3: n= 24 direct surface flow measurements).

Unfortunately, logger breakdowns occurred during the runoff measurement probablyly because of the frequently wet conditions in the forest studied. Data gaps were closed by means of the hydrological modelling program TOPMODEL [*Beven et al.*, 1995] as described in Fleischbein *et al.* [2006].

Water contents of soils were recorded hourly by Frequency Domain Reflectrometry (FDR) sensors (ML2 Theta, Delta T Devices, Cambridge, UK)) in MC2. The sensors were installed in the O horizon and at 0.1 m, 0.2 m, 0.3 m, and 0.4 m depth in the mineral soil.

3.4 Water analyses

For all analyses, samples were filtered in the field laboratory (ashless white ribbon paper filters, pore size, 4-7 µm, Schleicher and Schuell, Dassel, Germany). In one aliquot pH was determined; another aliquot was stored frozen until exported to Germany.

Water samples were analyzed colorimetrically with a segmented continuous flow analyzer (SAN^{Plus}, Skalar Analytical B.V., Breda, the Netherlands) for concentrations of dissolved inorganic nitrogen (NH_4-N and NO_3-N + NO_2-N, hereafter referred to as NO_3-N), and total dissolved nitrogen (TDN) concentrations (after UV oxidation to NO_3). Total organic C (TOC) concentrations were determined with an automatic TOC analyzer (TOC-5050, Shimadzu, Duisburg, Germany), total dissolved phosphorus (TDP) and sulfur (TDS) concentrations with ICP-OES (Integra XMP, GBC Scientific Equipment, Dandenong, Victoria, Australia). Additionally, water samples were analyzed by flame atomic absorption spectroscopy (AAS) for concentrations of Ca, Mg, K, and Na. Concentrations of Al and Mn were determined with inductively-coupled plasma-mass spectroscopy (ICP-MS, VG PlasmaQuad PG2 Turbo Plus, VG Elemental, Thermo Fisher Scientific, Waltham, Massachusetts, U.S.A.).

3.5 Soil analyses

The following soil properties were determined: soil pH in water (soil:solution ratio 1:2.5 v/v) with a standard pH electrode (Orion U402-S7, Thermo Fisher Scientific, Waltham, Massachusetts, U.S.A.), total C, N, and S concentrations with a CHNS-analyzer (Vario EL, Elementar Analysensysteme, Hanau, Germany), effective cation-exchange capacity (ECEC) by extraction with 1 M NH_4NO_3, base saturation (BS) by calculating the proportion of charge equivalent of extractable Ca + K + Mg + Na of the ECEC. Total P and metal concentrations in the mineral soil samples were determined by digestion with concentrated HNO_3/HF (4:1 v/v) and in the organic samples by digestion with concentrated HNO_3 under pressure. The concentrations of P in the extracts were determined with inductively-coupled plasma-atomic emission spectrometry (ICP-AES, GBC Integra XMP). Metal concentrations in the extracts were determined as described under 'water analyses'.

3.6 Calculations and statistical evaluation

Carbon, N, P, S, Ca, Mg, K, Na, Al, and Mn fluxes were calculated for rainfall, throughfall, and stemflow by multiplying the respective annual volume-weighted mean (VWM) concentrations with the annual water fluxes. Carbon, N, P, S, Ca, Mg, K, Na, Al, and Mn fluxes with surface runoff were calculated by multiplying flow-weighted mean (FWM) concentrations with the measured or modeled annual surface runoff and referring the annual flux to the surface area of the catchments.

The concentrations of DON were calculated as difference between those of TDN and NH_4-N + NO_3-N. Some samples had concentrations below the detection limit of the analytical methods (0.075 mg l^{-1} for N, 0.2 mg l^{-1} for P, 0.3 mg l^{-1} for S, 0.001 mg l^{-1} for Ca, Mg, K, and Na, 0.005 mg l^{-1} for Al, and 0.002 mg l^{-1} for Mn). For calculation purposes, values below the detection limit were set to zero. Thus, our annual means underestimate the real concentration of chemical constituents and mean concentrations can be smaller than the detection limit. Water fluxes in the soil were not quantified. We therefore could not calculate VWM values and instead show the median in the Tables.

For description of the water flow regime, stream water samples collected in weekly interval at our weirs were grouped into five flow classes representing the type of discharge event at the time of sampling. Flow classes were defined by their relation to the 5-year mean discharge for each catchment as modelled with TOPMODEL [*Fleischbein et al.*, 2006]. Discharge of flow class "superdry" was defined as less than 25%, flow class "baseflow" as between 25% and 50%, and flow class "intermediate" as ranging between 50% and 200% of the 5-year mean discharge of a catchment (Figure D-1).

High discharge periods were further divided into the flow classes "stormflow" and "lateral flow". Stormflow was defined as occurring if the discharge was more than double the 5-year mean. Lateral flow met the same criterion but in addition was associated with an at least four times greater 12-hour throughfall than the 5-year mean. The throughfall criterion substitutes the soil water saturation criterion of >85% of the maximum soil water content of the O horizon and the uppermost 0.2 m of the mineral soil (but not necessarily for the 0.2-0.4 m depth layer, Figure D-2). This substitution was necessary because of frequent breakdowns of the FDR data logger in MC2 and no available FDR data in MC1 and MC3. The throughfall criterion proved to be equally valid in 100% of the lateral flow events where both throughfall and soil water content data were available (>50% of all events in MC2, Figure D-2).

Significant differences among flow classes were tested with a *post-hoc* test (Games Howell). Statistical analyses were performed with SPSS 13.0 for Windows XP (SPSS Inc. 2000, http://www.spss.com/).

Based on the assumption that the flow class "superdry" with its lowest discharge rate represents the concentrations of chemical constituents in groundwater, we calculated a "flow class effect" (FE) to estimate the contribution of each flow class to the concentrations of chemical constituents in stream water (equation (7)).

$$FE_{[Afc]} = (\text{conc } [A]_{fc}) - (\text{conc}[A]_{SD}\ q_{sd}\ q_{fc}^{-1}), \tag{7}$$

where A is the specific chemical constituent, fc is the respective flow class, q_{sd} the discharge of the flow class "superdry" [litre s^{-1}], q_{fc} the discharge of the respective flow class [litre second^{-1}], conc[A] the concentration of A in SD ("superdry flow") or another flow class (fc) [mg litre^{-1}]). FE [mg litre^{-1}] was calculated as the 5-year mean (April 1998- April 2003) and expresses the difference between the measured concentration for fc (left part of term of the right side in Equation (7)) and the hypothetical concentration the chemical constituent A would have if only diluted by the additional discharge of the flow class fc (right part of the right side in equation (7)).

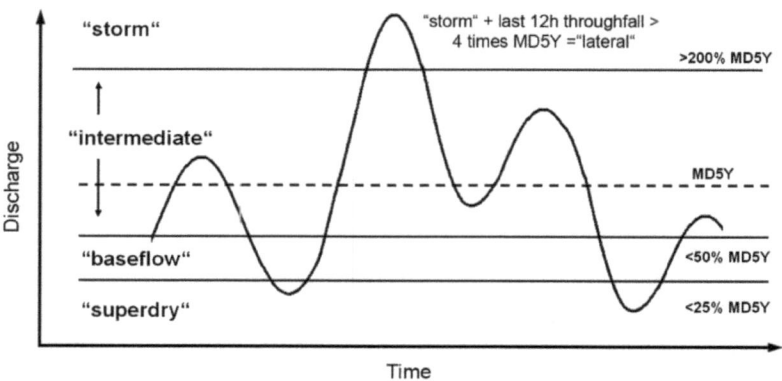

Figure D-1: Schematic sketch of the five flow classes (MD5Y= mean discharge over five years).

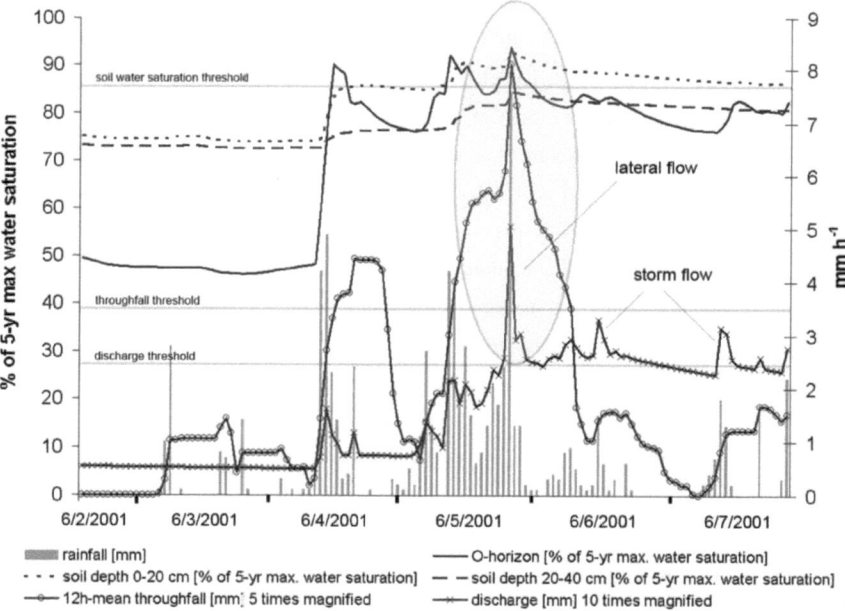

Figure D-2: Illustration of the suitability of the throughfall criterion to represent the soil water content criterion to separate lateral flow from stormflow conditions by the example of an event between 6/2/2001 and 6/7/2001. When discharge threshold and either throughfall threshold or soil water saturation threshold are crossed, lateral flow conditions apply.

To estimate the contribution of different flow classes to total element export we developed a simple new model, in which we classified hourly discharge [obtained with TOPMODEL, *Fleischbein et al.*, 2006] into our five flow classes. Then, we multiplied the cumulative discharge for a given flow class during the monitored 5-year period with the mean concentration of each of the chemical constituents studied in the same flow class. Finally, the export rates of each flow class and each chemical constituent were summed to give a total 5-year element export for each catchment. To validate this new approach, we compared its results with conventionally derived results from flow-weighted means and cumulative discharge.

3.7 Modelling of dissolved metal speciation

To further investigate into the controls of metal solubility during the various flow conditions, we calculated humic complexation of metals in water samples by the speciation model VisualMinteQ v. 2.53 [*Gustafsson et al.*, 2001]. We used the Nica-Donnan model sub-routine with the active DOM to DOC ratio set to 1.4 and assuming 100% of active DOM to be fulvic acids. Furthermore, we assumed concentrations of TOC to be equal to concentrations of DOC although our samples were filtered through 4-7 μm pores. We tested the stability of the observed trends in metal complexation over the five flow classes by assuming that only 50% of TOC was DOC and did not observe significant changes.

4. Results and Discussion

4.1 Hypothesis 1: The concentrations of chemical constituents in stream water of steep, forested catchments are related to discharge levels in a way that is specific for each chemical constituents

Flow regime. The mean annual discharge for the five monitored years modelled with TOPMODEL was 958 mm for MC1 (716-1168 mm), 1101 mm for MC2 (1017-1197 mm), and 1032 mm for MC3 (878-1137 mm). During the monitored period, the least frequently observed flow regime was peak discharge (sum of our flow classes "stormflow" and "lateral flow", which might be similar to what is termed "stormflow" in other studies), ranging from 23 samples (8.7%) for MC2 to 38 samples (14.4%) for

Table D-1: Occurrence and seasonal distribution of measured and modelled discharge levels over the five flow classes of three small streams draining microcatchments (MC1-3) under montane rain forest in southern Ecuador (April 1998- April 2003).

	Sampled distribution Samples (n=263)			Modelled distribution[a] Hours (n=44206)			Sampled distribution %			Modelled distribution %			Seasonal distribution (mean MC1-3) Hours per season				% per season[b]			
Flow class	MC1	MC2	MC3	MC1	MC2	MC3	MC1	MC2	MC3	MC1	MC2	MC3	1[c]	2[c]	3[c]	4[c]	1[c]	2[c]	3[c]	4[c]
Superdry	53	8	27	1650	2532	3139	20	3.0	10	3.7	5.7	7.1	1621	153	11	654	67	6.3	0.5	27
Baseflow	77	62	75	13303	12126	10000	29	24	29	30	27	23	3755	1336	974	5745	32	11	8.2	49
Intermediate	95	170	135	24029	25352	26365	36	65	51	54	57	60	5375	6725	8546	4602	21	27	34	18
Storm	30	14	19	4440	3144	3968	11	5.3	7.2	10	7.1	9.0	46	2440	1338	27	1.2	63	35	0.7
Lateral	8	9	7	784	1052	734	3.0	3.4	2.7	1.8	2.4	1.7	25	648	172	12	2.9	76	20	1.4

[a] TOPMODEL, Fleischbein et al. (2006) and own results for 2002/2003
[b] percentage values indicate the distribution of a class among seasons
[c] seasons; 1: Januar to March; 2: April to June; 3: July to September; 4: October to December

MC3 (ranges of "storm": 5.3%-11.4% and "lateral": 2.7%-3.4%, Table D-1). Low discharge conditions (consisting of the classes "superdry" and "baseflow") occurred more frequently, ranging from 70 samples (26.6%) for MC2 to 130 samples (49.4%) for MC1. The flow class "intermediate", representing the majority of discharge events, was observed least frequently in MC1 (36% of the samples) and most frequently in MC2 (65%, Table D-1).

The modelled distribution of flow events among our five flow classes was similar to that of the water level record taken during our weekly sampling. The most pronounced deviation of the modelled from the sampled distribution was that superdry conditions were more frequently sampled in MC1 where, in turn, intermediate flow was under-represented by our sampling compared with the model (Table D-1).

Approximately three-fourths of the peak discharge events (i.e., stormflow+lateral flow) occurred between April and June (Table D-1), receiving 39% (982mm) of the mean annual rainfall (2520mm), which is more than twice the amount of the January-March (472mm) and October-December (474mm) periods (Table D-1). Almost all of the remaining stormflow events occurred between July and September (952mm, Table D-1). Thus, the hydrological year usually was stormflow-dominated from April to September and baseflow-dominated from October to March.

C, N, P and S. Carbon, N, P, and S form two groups with respect to the relation between discharge and concentration. The first group, P and S, had maximum concentrations in stream water when discharge was low (Figure D-3e1 and f1). In contrast, the second group, consisting of TOC and the N species, had maximum concentrations when discharge was high (Figure D-3a1, d1). The second group is subdivided into TOC, NH_4-N, DON with greater and NO_3-N with smaller concentrations (except in MC2) during lateral flow, when the soil is waterlogged, than during stormflow (Figure D-3). In particular, TOC and DON more than tripled their concentrations under "lateral flow" compared with "stormflow" conditions (Fig. D-3).

To assess possible dilution/concentration effects, we calculated the "flow class effect" (FE, values for q of the FE calculation are shown in Table D-2). If FE of a chemical constituent was zero for each flow class, its concentration in stream water was only explained by dilution. This was only true for P (Table D-3, Figure D-4). "Flow class effects" >0 indicate changes in the concentrations of a chemical constituent that cannot be explained by dilution alone. The FE of TOC, the N species and S increased continuously with increasing discharge (Figure D-4, SD-SF).

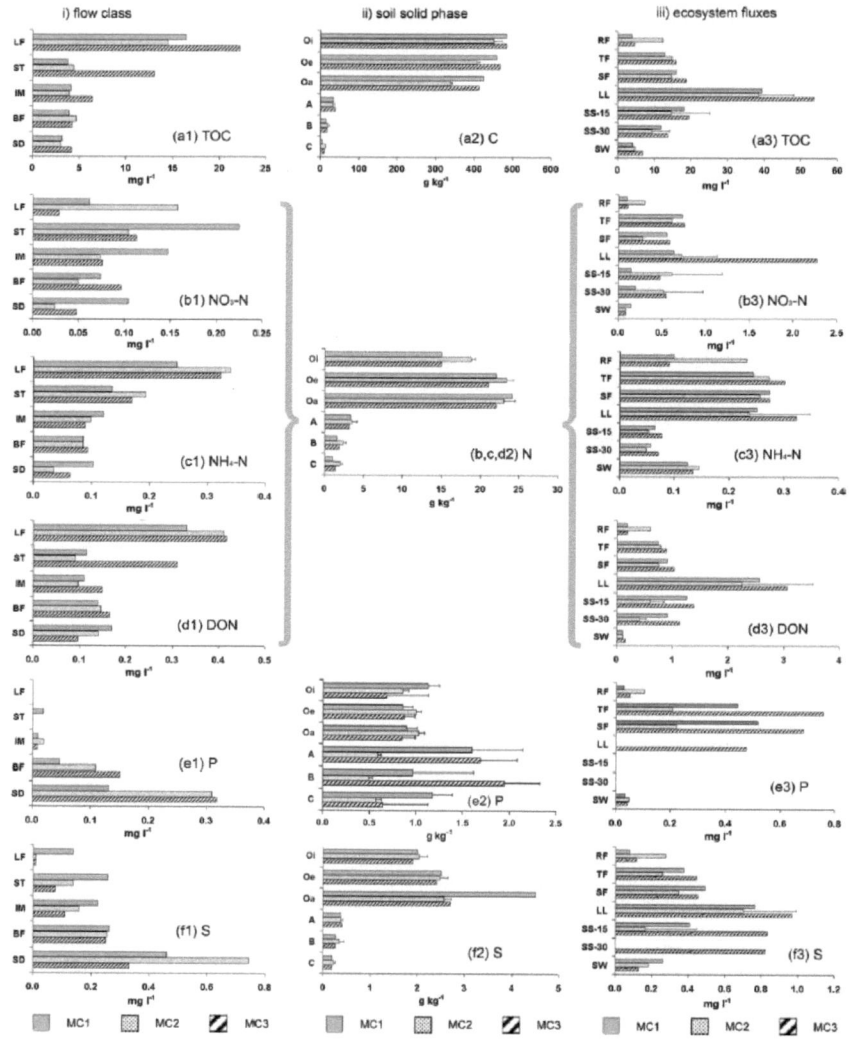

Figure D-3: Mean concentrations of TOC, N, P, and S in (i.) flow class, (ii.) soil solid phase, and (iii.) ecosystem flux between April 1998 and April 2003 in the three microcatchments (MC 1-3). Abbreviations: lateral flow (LF), storm flow (ST), intermediate (IM), base flow (BF), superdry (SD), rainfall (RF), throughfall (TF), stemflow (SF), litter leachate (LL), soil solutions at 0.15m (SS-15) and 0.30m (SS-30) and streamwater (SW). Error bars indicate standard error.

If the soil was saturated before the higher flow event (flow class "lateral flow), the FEs of TOC, NH_4-N, and DON increased further, while those of NO_3-N and S markedly decreased (Figure D-4, LF). Given the marked differences in the concentrations of chemical constituents and FEs among low and high discharge classes, it does not seem likely that the event water displaced groundwater as suggested for moderately steep catchments [*Genereux*, 2004]. Instead, our results can be explained by the mixing of chemically different event water with pre-event water [*Brown et al.*, 1999; *Goller et al.*, 2005; *Schellekens et al.*, 2004].

The TOC, NO_3-N, NH_4-N, and DON concentrations in stream water are best explained by the assumption that rapid, near-surface flow carries elevated concentrations of these chemical constituents to the stream where the solutions mix with groundwater-fed baseflow [*Schellekens et al.*, 2004; *Tromp-van Meerveld and McDonnell*, 2006]. Goller et al. [2005] have shown with a $\delta^{18}O$ approach that at our study site rapid near-surface flow in soil only occurs if the soil is saturated prior to the rainstorm.

Besides the triggering of lateral flow during storm conditions, water saturation in soil reduces mineralization [*Likens et al.*, 2002]. This would result in reduced N, P, and S release and reduced nitrification and thus explains the smaller S and NO_3-N concentrations under "lateral flow" compared to "storm flow" conditions. Furthermore, SO_4 and NO_3 are rapidly leached in the first discharge event after a drier period as a consequence of their mobility in soil.

Table D-2: Mean 5-year discharge for the different flow classes and MCs, used as q in the flow class effect (FE) calculation

Flow class	5-year mean discharge [l s^{-1}]		
	MC1	MC2	MC3
Superdry	0.44	0.64	0.60
Baseflow	0.83	1.34	1.36
Intermediate	2.59	3.42	3.13
Storm	7.19	10.86	12.21
Lateral	9.83	12.29	28.83

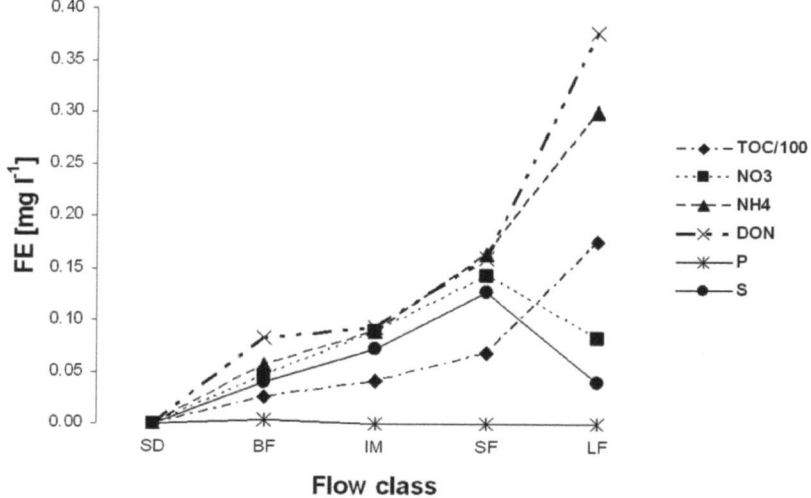

Figure D-4: Estimated influence of a flow class on the concentrations of chemical constituents in stream water ("flow class effect", FE). Abbreviations as in Figure D-3. Note: FE of TOC reduced by a factor of 100.

Metals. The monitored metals can be classified into three groups in terms of their concentration-discharge level behaviour. The concentrations in stream water of the first group, K, Ca, and Mg, tended to increase continuously with increasing discharge levels (Figure D-5, a1-c1). The differences among the flow classes were not significant except that K concentrations in superdry flow were significantly less than in all other flow classes (Table D-4). Aluminium and Mn formed the second group with almost constant concentrations during most discharge levels except for "lateral flow" where their concentrations in stream water increased by a factor of up to 20 (Figure D-5, e1 and f1, Table D-4). In MC3, this effect was also observed for the flow class "storm". In contrast to all other monitored metals, Na - the only representative of the third group of metals - showed a negative relation with discharge levels. The decrease in Na concentrations with increasing discharge level was significant in all three MCs (Figure D-5, d1, Table D-4).

The FEs of all monitored metals increased continuously with increasing discharge (Figure D-6, SD-SF). If the soil was saturated before the higher flow event (flow class "lateral flow"), the FEs of Al and Mn increased markedly while those of K, Ca, and Mg increased almost linearly through all flow classes from superdry to lateral (Figure D-6, LF). In contrast to all other metals, the FE of Na decreased during lateral flow conditions (Figure D-6, LF). These results demonstrate that the differences in metal concentrations among flow classes cannot be explained by dilution alone.

Table D-3: Mean concentration of C, N, P, and S in stream water and flow class effect (FE) of the different flow classes for the three streams. Different lower case letters indicate significant differences among flow classes (Games-Howell, p<0.05)

	stream	Superdry conc.	Baseflow conc.	FE	Intermediate conc.	FE	Storm conc.	FE	Lateral conc.	FE
Number of samples	MC1	53	77	-	95	-	30	-	8	-
	MC2	8	62	-	170	-	14	-	9	-
	MC3	27	75	-	135	-	19	-	7	-
					mg l^{-1}					
TOC	MC1	3.1b	3.9b	2.2	4.1b	3.6	3.8b	3.6	16.4a	16.2
	MC2	3.1b	4.6b	3.2	3.9b	3.3	4.4b	4.2	14.4a	14.3
	MC3	4.2b	4.2b	2.4	6.4b	5.6	13.1a	12.9	22.1a	22.1
NO$_3$-N	MC1	0.10b	0.07b	0.02	0.15b	0.13	0.22a	0.22	0.06b	0.06
	MC2	0.02b	0.05b	0.04	0.07b	0.07	0.10a	0.10	0.16a	0.16
	MC3	0.05b	0.10a	0.08	0.08a	0.07	0.11a	0.11	0.03c	0.03
NH$_4$-N	MC1	0.10a	0.09a	0.03	0.12a	0.10	0.14a	0.13	0.25a	0.24
	MC2	0.04a	0.09a	0.07	0.10a	0.09	0.19a	0.19	0.34a	0.34
	MC3	0.06a	0.09a	0.07	0.09a	0.08	0.17a	0.17	0.32a	0.32
DON	MC1	0.17a	0.14a	0.05	0.11a	0.08	0.12a	0.10	0.33a	0.32
	MC2	0.14a	0.15a	0.08	0.10a	0.07	0.09a	0.08	0.41a	0.40
	MC3	0.10c	0.17b	0.12	0.15b	0.13	0.31a	0.30	0.42a	0.41
P	MC1	0.13a	0.05a	0	0.01b	0	0.02b	0	0c	0
	MC2	0.31a	0.11a	0	0.02a	0	0a	0	0a	0
	MC3	0.32a	0.15a	0.01	0.01b	0	0b	0	0c	0
S	MC1	0.46a	0.26a	0.02	0.22b	0.15	0.26b	0.23	0.14c	0.12
	MC2	0.74a	0.26a	0	0.16b	0.02	0.14b	0.09	0.01c	0
	MC3	0.33a	0.25a	0.10	0.11b	0.05	0.08b	0.06	0.01c	0

4.2 Hypothesis 2: The depth of water flow in soil determines the concentration of chemical constituents in stream water

C, N, P, and S. Carbon, N, P, and S accumulated markedly in the organic layer compared with the mineral soil (up to >100 times for P and S, sixfold for C, fourfold for N except in MC2, Figure D-3, a2-f2). Concentrations of C and N tended to be greater in the Oi and Oe horizons, those of P and S were greatest in the Oa horizon.

The concentrations of all chemical constituents increased in throughfall and stemflow relative to rainfall after their passage through the canopy (Figure D-3 a3-f3). This is attributable to dry deposition and canopy leaching [*Parker*, 1983]. The canopy is a source of all the chemical constituents studied, because the increase in concentrations is more pronounced than can be explained by evaporation and resulting concentration [*Goller et al.*, 2006].

The median of the concentrations of TOC, NO_3-N, and DON in litter leachate was more than double those in throughfall or stemflow, the median of the S concentrations increased by about 50% and that of P decreased to below the detection limit of 0.2 mg l^{-1}, except for MC3. The NH_4-N concentrations did not change between throughfall (VWM) and litter leachate (median). The medians of the concentrations of all chemical constituents in soil solution decreased consistently compared to litter leachate. Between 0.15 and 0.30 m mineral soil depth the medians of the concentrations of NH_4-N and NO_3-N did not change; those of TOC, DON, S, and P decreased with increasing depth. The FWM concentrations of TOC, NO_3-N, and DON in stream water were less than their medians in soil solution and those of NH_4-N, P, and S were greater except for S in MC3. Considering the depth distribution of C, N, P, and S in the soil solid phase and soil solution suggests that the deeper water percolates through the soil the smaller the concentrations of chemical constituents draining to the stream should be. This was indeed the case for TOC and N species where "lateral flow" and "storm flow" had the greatest concentrations among all flow classes, but not for P and S.

Although P accumulated in the topsoil (Figure D-3e2), we did not detect P in the litter leachate (Figure D-3e3). This suggests that P was completely retained by the vegetation because the organic layer is densely rooted and P might limit growth [*Wilcke et al.*, 2002]. The only exception was MC3, where the pH was less acid and therefore provided better conditions for microorganisms, thus resulting in faster mineralization, which released more P than in the other two catchments [*Wilcke et al.*, 2001; 2002].

However, except in MC3, all P was retained in the near-surface mineral soil presumably by sorption so that flushing of the surface soil did not result in P transport to the stream.

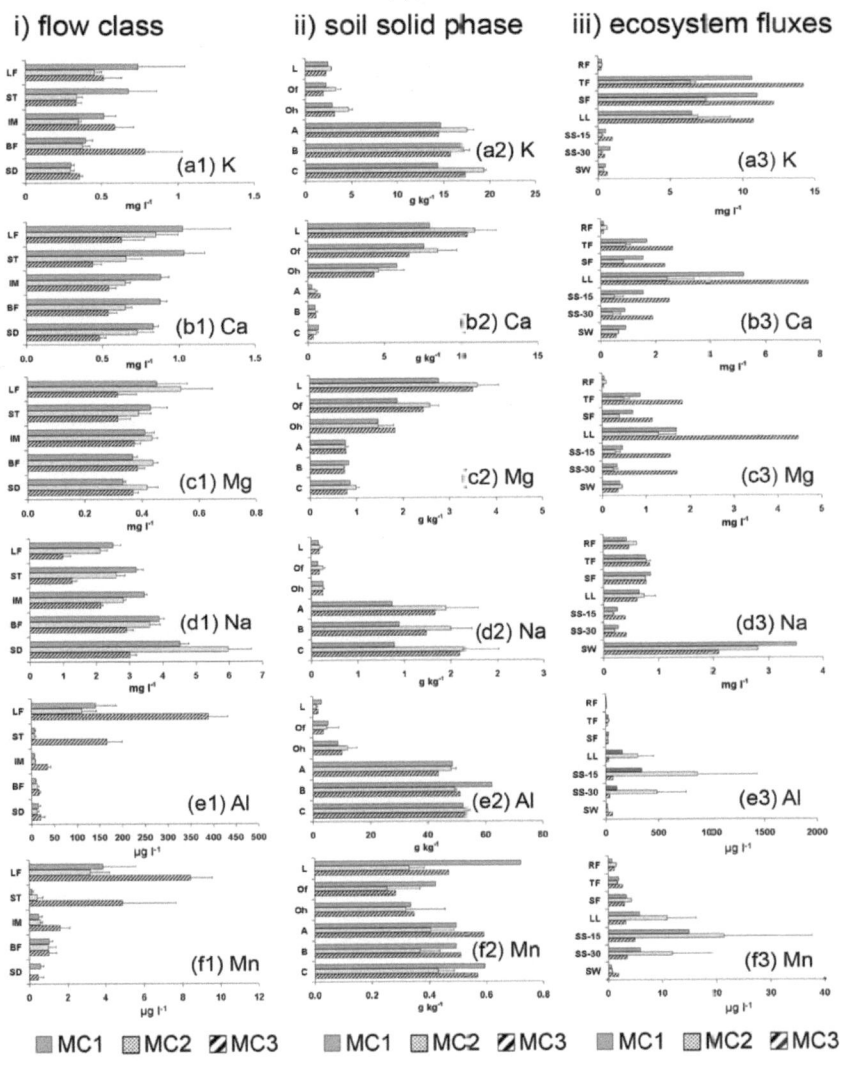

Figure D-5: Mean concentrations of K, Ca, Mg, Na, Al, and Mn in (i.) flow class, (ii.) soil solid phase, and (iii.) ecosystem flux between April 1998 and April 2003 in the three microcatchments (MC 1-3). Abbreviations as in Figure D-3. Error bars indicate standard errors.

Table D-4: Mean concentration of K, Ca, Mg, Na, Al, and Mn in stream water and flow class effect (FE) of the different flow classes for the three streams. Different lower case letters indicate significant differences among flow classes (Games-Howell, p<0.05).

	stream	Superdry conc.	Baseflow conc.	FE	Intermediate conc.	FE	Storm conc.	FE	Lateral conc.	FE
Number of samples	MC1	53	77	-	95	-	30	-	8	-
	MC2	8	62	-	170	-	14	-	9	-
	MC3	27	75	-	135	-	19	-	7	-
		\multicolumn{9}{c}{mg l^{-1}}
K	MC1	0.30b	0.40a	0.24	0.52a	0.47	0.68a	0.66	0.74a	0.73
	MC2	0.29b	0.38a	0.24	0.35a	0.29	0.34a	0.32	0.46a	0.44
	MC3	0.35a	0.78a	0.63	0.59a	0.52	0.33a	0.31	0.51a	0.50
Ca	MC1	0.83a	0.88a	0.44	0.88a	0.74	1.0a	0.99	1.0a	0.99
	MC2	0.73a	0.65a	0.30	0.65a	0.51	0.65a	0.61	0.85a	0.81
	MC3	0.48a	0.54a	0.32	0.54a	0.45	0.44a	0.41	0.62a	0.61
Mg	MC1	0.33a	0.37a	0.19	0.41a	0.36	0.43a	0.40	0.45a	0.44
	MC2	0.42a	0.44a	0.24	0.44a	0.36	0.39a	0.36	0.54a	0.52
	MC3	0.37a	0.38a	0.22	0.37a	0.30	0.32a	0.30	0.32a	0.31
Na	MC1	4.5a	3.9a	1.5	3.4b	2.7	3.2b	2.9	2.5c	2.3
	MC2	6.0a	3.6b	0.75	2.8b	1.7	2.6b	2.3	2.1c	1.8
	MC3	3.0a	2.9a	1.6	2.1b	1.6	1.3c	1.1	1.0c	0.94
		\multicolumn{9}{c}{µg l^{-1}}
Al	MC1	14b	8.4b	0.89	5.9b	3.5	7.0b	6.1	141a	140
	MC2	12b	13b	7.3	7.7b	5.5	6.7b	6.0	110a	110
	MC3	19c	16c	7.7	35c	31	166b	165	389a	389
Mn	MC1	0.57a	1.0a	0.73	0.48a	0.38	0.13a	0.10	3.9a	3.83
	MC2	0c	0.97b	0	0.57b	0.19	0.42b	0.30	3.2a	3.1
	MC3	0.47b	1.0b	0.79	1.6b	1.5	4.9a	4.9	8.4a	8.4

Although P accumulated in the topsoil (Figure D-3e2), we did not detect P in the litter leachate (Figure D-3e3). This suggests that P was completely retained by the vegetation because the organic layer is densely rooted and P might limit growth [*Wilcke et al.*, 2002]. The only exception was MC3, where the pH was less acid and therefore provided better conditions for microorganisms, thus resulting in faster mineralization, which released more P than in the other two catchments [*Wilcke et al.*, 2001; 2002]. However, except in MC3, all P was retained in the near-surface mineral soil presumably by sorption so that flushing of the surface soil did not result in P transport to the stream. The large concentrations of P in "superdry" and "baseflow" therefore must result from weathering of the mineral subsoil. Thus, there seem to be two decoupled P cycles in the vegetation and organic layer and the lower subsoil. The upper part of this cycle seems to be almost completely closed so that no P losses to the mineral soil occur.

The greatest S concentrations in "superdry" and "baseflow" also suggest that S was released from the mineral soil by weathering. The weathering probably also included reoxidation of previously reduced S during waterlogging [*Likens et al.*, 2002]. However, at the same time large concentrations in litter leachate indicated that retention by the vegetation was not as complete as for P. Therefore, we observed measurable S concentrations during stormflow (Figure D-4).

These results support our second hypothesis for TOC and TDN concentrations. However, for N speciation and total concentrations of P and S there were additional controls of the stream water concentrations during different discharge levels. These controls include biological processes such as microbial nitrification and nutrient uptake by soil organisms and plants and mineral weathering in the subsoil.

Metals. All metals had their greatest concentrations in the litter leachates below the organic layer except for Na where the greatest concentrations occurred in the stream water (Figure D-5, a3, b3, c3).

The greatest Ca and Mg concentrations in the soil solid phase occurred in the O horizons (Figure D-5, b2, c2, Table A-1). Total concentrations of K peaked in the subsoil, but exchangeable K concentrations were greatest in the O horizons if it is assumed that all K in the O horizons is exchangeable (Figure D-5, a2, a3, Table A-1). Potassium is strongly cycled by the vegetation and therefore abundant in plant-available form in the O horizons but less mobile in the mineral soil where it is fixed by the abundant illite present in the study soils [*Schrumpf et al.*, 2001, Figure D-5, a3]. All elements of this group were increasingly mobilized by decreasing pH with decreasing flow depth in soil (Figure D-7, a-c). Decreasing flow depth in soil also caused increased TOC concentrations in soil water, leading to increasing organo-complexation (or ion association to satisfy the electroneutrality requirement).

Table D-5: Flow-weighted mean (fwm) concentration, "traditional" (fwm-based) export, modelled export, and difference between traditional and modelled export of C, N, P, and S of the three microcatchments between April 1998 and April 2003.

	Catchment	Discharge mean (range) mm	Concentration (fwm) mg l⁻¹	Export ("traditional")	Export (modelled)	Literature data (Tropics) kg ha⁻¹ year⁻¹	Difference (between modelled and traditional)	%
TOC	MC1	958 (716-1168)	4.0	39	45		6.5	17
	MC2	1101 (1017-1197)	4.8	53	63	11.3[d]	9.7	18
	MC3	1032 (878-1137)	6.8	71	92		21	29
NO₃-N	MC1	958 (716-1168)	0.14	1.4	1.4	1.39-6.10[a]	0.02	1.5
	MC2	1101 (1017-1197)	0.08	0.88	1.0	0.23[c]	0.12	13
	MC3	1032 (878-1137)	0.09	0.89	0,87	0.1[d]	-0.02	2.3
NH₄-N	MC1	958 (716-1168)	0.12	1.2	1.2	0.3-0.7[a]	0.02	1.3
	MC2	1101 (1017-1197)	0.14	1.6	1.7	0.34[c]	0.10	6.2
	MC3	1032 (878-1137)	0.13	1.4	1.3	0.1[d]	-0.09	6.7
DON	MC1	958 (716-1168)	0.12	1.1	1.2	1.5-4.8[a]	0.11	10
	MC2	1101 (1017-1197)	0.12	1.4	1.6	3.79[c]	0.26	19
	MC3	1032 (878-1137)	0.17	1.8	2.2	0.5[d]	0.36	20
P	MC1	958 (716-1168)	0.03	0.31	0.15	0.03-0.08[b]	-0.16	51
	MC2	1101 (1017-1197)	0.05	0.54	0.30		-0.24	46
	MC3	1032 (878-1137)	0.04	0.45	0.23	0.79[c]	-0.20	46
S	MC1	958 (716-1168)	0.26	2.5	2.1	14-24[b]	-0.36	14
	MC2	1101 (1017-1197)	0.18	2.0	1.7		0.29	14
	MC3	1032 (878-1137)	0.13	1.3	1.1	3.38[c]	-0.19	15

[a] Lewis et al. (1999), overview of various studies, here only catchments of comparable size and elevation selected; [b] McDowell and Asbury (1994); [c] Liu et al. (2003); [d] Möller et al. (2005)

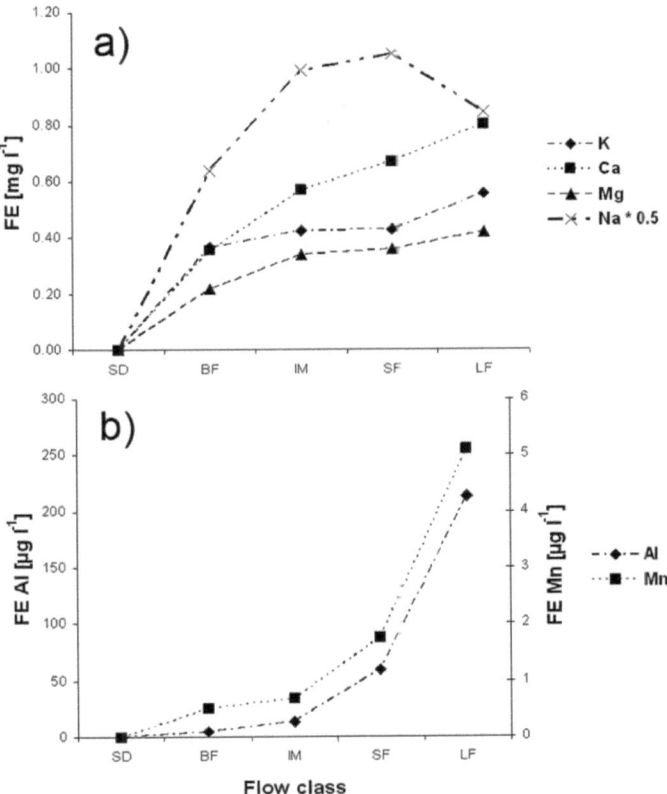

Figure D-6: Estimated influence of a flow class on the concentrations of chemical constituents in stream water ("flow class effect", FE) for a) base metals and b) Al and Mn. Abbreviations as in Figure D-3. Note: FE of Na reduced by a factor of 2.

In our MinteQ-based speciation calculations all metals showed increasing complexation with increasing discharge and therefore increasing TOC concentrations (Figure D-7). Thus, organo-complexation was a major driver of metal mobilization in the study soil. Amounts of K bound to TOC were negligible compared to those of Ca and Mg (Figure D-7, a-c). The vertical variation in pH and DOM concentrations also explained the observed FEs (Figure D-6a).

Table D-6: Flow-weighted mean (fwm) concentration, "conventional" (fwm-based) export, modelled export, and difference between modelled and measured export of K, Ca, Mg, Na, Al, and Mn of the three microcatchments between April 1998 and April 2003.

	Catchment	Discharge mean (range) mm	Concentration (fwm)	Export ("traditional")	Export (modelled)	Literature data (Tropics) kg ha⁻¹ year⁻¹	Difference (between modelled and traditional)	%
K	MC1	958 (716-1168)	0.52 mg l⁻¹	5.0	5.2	3.5-5.8^a	0.20	4
	MC2	1101 (1017-1197)	0.35 mg l⁻¹	3.9	4.1	7.0-11^b	0.25	7
	MC3	1032 (878-1137)	0.63 mg l⁻¹	6.5	5.5	4.9-17c 17d	-0.98	15
Ca	MC1	958 (716-1168)	0.90 mg l⁻¹	8.7	8.6	5.2-13.9^a	-0.06	0.7
	MC2	1101 (1017-1197)	0.64 mg l⁻¹	7.0	7.7	3.6-7.0^b	0.68	10
	MC3	1032 (878-1137)	0.56 mg l⁻¹	5.8	5.4	44-96^c 34^d	-0.44	8
Mg	MC1	958 (716-1168)	0.40 mg l⁻¹	3.8	3.8	2.3-5.5^a	0.02	0.5
	MC2	1101 (1017-1197)	0.44 mg l⁻¹	4.9	5.0	3.6-6.0^b	0.10	2
	MC3	1032 (878-1137)	0.35 mg l⁻¹	3.6	3.7	28-63^c 8.7^d	0.10	3
Na	MC1	958 (716-1168)	3.5 mg l⁻¹	34	31	2.7-3.5^b	-2.7	8
	MC2	1101 (1017-1197)	2.8 mg l⁻¹	31	31	96-161^c 1.2^d	0.96	3
	MC3	1032 (878-1137)	2.1 mg l⁻¹	22	20		-1.6	7
Al	MC1	958 (716-1168)	12 μg l⁻¹	0.12	0.16		0.04	31
	MC2	1101 (1017-1197)	15 μg l⁻¹	0.17	0.25	---	0.08	49
	MC3	1032 (878-1137)	61 μg l⁻¹	0.62	0.92		0.30	48
Mn	MC1	958 (716-1168)	0.59 μg l⁻¹	0.006	0.007		0.001	15
	MC2	1101 (1017-1197)	0.74 μg l⁻¹	0.008	0.011	---	0.003	30
	MC3	1032 (878-1137)	1.9 μg l⁻¹	0.029	0.020		0.010	50

^{a} Stoorvogel et al. (1997); ^{b} Yusop et al. (2006); ^{c} McDowell and Asbury (1994); ^{d} Liu et al. (2003)

Total concentrations of Al were greatest in the mineral soil, while total concentrations of Mn were relatively evenly distributed among all soil horizons (Figure D-5, e2, f2). Although considerable concentrations of Al and Mn were found in the litter leachates and the mineral soil (Figure D-5, e3 and f3), caused by pH values <5 especially in MC2 (Figure D-7), great concentrations of Al and Mn in stream water only occurred during storm and lateral flow conditions (Figure D-5, e1 and f1). The degree of organo-complexation of Al was almost constant and consistently >90% in all flow classes), while increasing TOC concentrations caused increasing complexation of Mn during "storm" and "lateral" flow conditions (Figure D-7, a-c). This observation is corroborated by the FEs of Al and Mn which increased markedly during low pH and TOC- rich "storm" and "lateral" flow conditions, while low FEs for Al and Mn during the flow classes "superdry" to "intermediate" indicated that stream water concentrations of Al and Mn were mainly controlled by dilution effects during low discharge conditions (Figure D-6, b). In MC3, stream water had a pH <5 during "lateral" flow conditions possibly decreasing the stability of the organo-metal complexes (Figure D-7). Sodium was almost absent in the O horizons of the ecosystem but abundant in the mineral soil (Figure D-5, d2). Sodium had its greatest concentrations in stream water (Figure D-5, d3). The Na concentrations in stream water decreased with decreasing water flow depth in soil (Figure D-5, d1). This reflects the vertical distribution of Na concentrations in soil. Furthermore, decreasing pH with decreasing water flow depth in soil might have enhanced the replacement of Na at the cation exchange sites (Figure D-7, a-c). Organo-complexation of Na was negligible except for the flow class "lateral flow" where about 10% of Na was bound to TOC (Figure D-7, a-c). The additional Na mobilization by organo complexation during lateral flow was not strong enough to compensate for the small Na concentrations in the O horizons as indicated by the decreasing FE for this flow class. This supports our assumption that considerable quantities of the event water never took the flow path through the Na-rich mineral soil horizons (Figure D-6a).

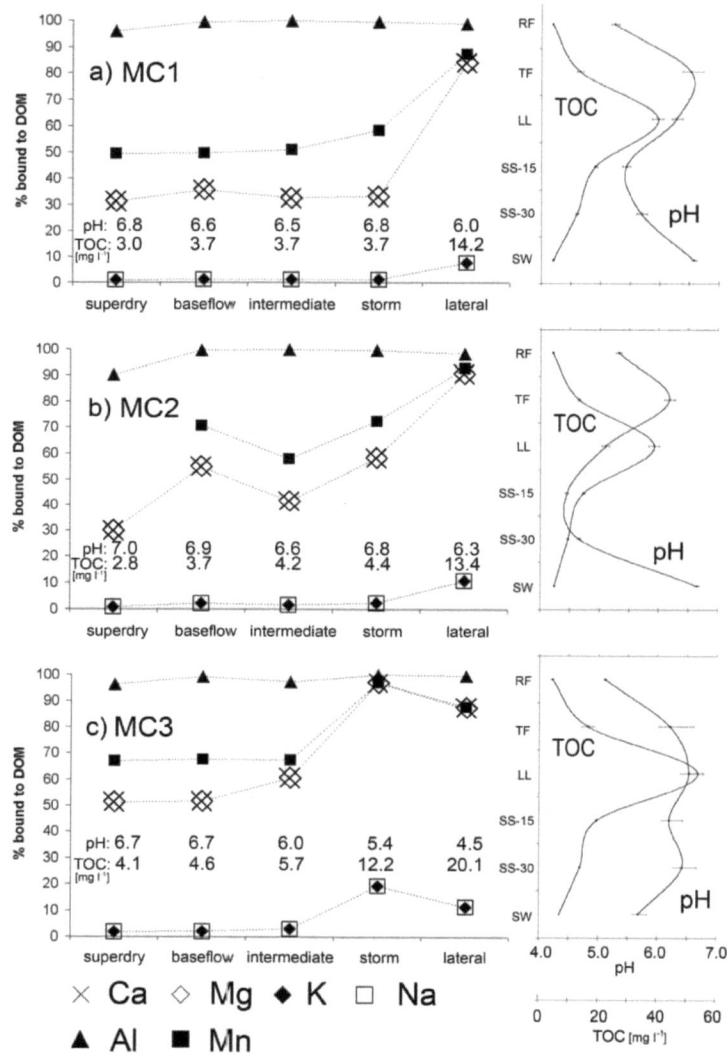

Figure D-7: Mean contributions of organo-metal complexes to total dissolved metal concentrations in the five flow classes for the three microcatchments (MC1-MC3, left panels a-c, as computed by MinteQ in weekly resolution) and mean depth profile of pH and flow-weighted mean concentration of TOC in ecosystem fluxes (right) between April 1998 and April 2003. Given pH and TOC values in the complexation graph correspond to the indicated flow classes. Abbreviations as in Figure D-3. Error bars indicate standard errors for the TOC concentrations and minima and maxima for the pH. Standard errors of the contributions of organo-metal complexes to total dissolved metal concentrations are smaller than the symbols.

4.3 Hypothesis 3: Discharge level classification is a suitable tool to estimate the contribution of different flow regimes to element export from steep, forested catchments for long time series

Traditional approach. The mean annual discharge over the 5-year period was 958 mm (MC1), 1101mm (MC2), and 1032 mm (MC3). The annual export calculated as FWM concentration times annual flow (hereafter referred to as "traditional export calculation") ranged from 39-71 kg TOC, 0.9-1.4 kg NO_3-N, 1.2-1.6 kg NH_4-N, 1.1-1.8 kg DON, 0.3-0.5 kg P, 1.3- 2.5 kg S ha^{-1}, 3.9-6.5 kg K ha^{-1} $year^{-1}$, 5.8-8.7 kg Ca ha^{-1} $year^{-1}$, 3.6-5.0 kg Mg ha^{-1} $year^{-1}$, and 22-34 kg Na ha^{-1} $year^{-1}$ (Tables D-5 and D-6).

Export rates of NO_3-N and DON were small and of NH_4-N were large compared to other forested microcatchments in the tropics. Phosphorus export was much greater, S export much less than the scarce literature values (Table D-5). The export rates of K, Ca, and Mg ranged at the lower end of the literature data from other tropical catchments of comparable size; Na export rates were average (Table D-6). We were not able to compare the export rates of Al (0.12-0.62 kg ha^{-1} $year^{-1}$) and Mn (0.006-0.03 kg ha^{-1} $year^{-1}$) at our study site to export rates of other catchments because we did not find other data.

Differences between our study site and the literature are most likely attributable to the geologically younger bedrock of volcanic origin (Cretaceous and younger), for most studies in contrast to the geologically old, Palaeozoic bedrock at our study sites. We only found two reports on C, N, P, and S export of catchments in the tropics and subtropics where the parent rock was not volcanic. Liu [2003] studied montane catchments under subtropical cloud forest in the Ailao Mountains (2400-2600 m.a.s.l.) in southern China and Möller *et al.* [2005] a relatively small (3.3 ha) patch of natural forest growing on comparable soils to those of our study site in a catchment in northern Thailand at an elevation between 1350 and 1500m a.s.l. (Table D-5). A more detailed comparison of C, N, P, and S export to other tropical catchments is found in Goller *et al.* [2006].

An alternative approach to considering varying concentrations related to different discharge rates for cumulative element export is to sample disproportionately during storm flow events. Then, regressions of concentrations on discharge can be used to assign a specific concentration to each discharge rate. Export rates are calculated by multiplying this concentration with the discharged water volume which then can be summed to give an annual export [*Buffam et al.*, 2001; *Evans and Davies*, 1998].

However, in our data set, discharge rate explained an insufficient part of the variation in concentrations of all chemical constituents (R^2 = 0.01-0.24 for log-transformed data) although there were marked differences in the concentrations of the chemical constituents among the different flow classes. Therefore, we could not use this approach and decided to develop a novel model to quantify the individual contributions of the five different flow classes. This new approach was validated by comparing the sums of the export rates in individual flow classes with the export rates calculated with the traditional FWM-based method. Assuming the results of the FWM-based method as true values requires that the FWMs were obtained from a data set which comprised representative samples for all flow conditions, which was the case in our study (Table D-1).

Table D-7: Mean element export of C, N, P, and S by discharge in different flow classes (April 1998- April 2003). Percentage values indicate the fraction of total export for the corresponding flow class.

	Catchment	Superdry		Baseflow		Intermediate		Storm		Lateral	
		kg ha^{-1} year^{-1}	%	kg ha^{-1} year^{-1}	%	kg ha^{-1} year^{-1}	%	kg ha^{-1} year^{-1}	%	kg ha^{-1} year^{-1}	%
TOC	MC1	0.22	0.5	4.1	9.0	20	44	9.4	21	11	26
	MC2	0.39	0.6	5.6	8.9	24	39	10	16	23	36
	MC3	0.52	0.6	4.0	4.3	38	41	35	38	15	16
NO$_3$-N	MC1	0.007	0.5	0.08	5.5	0.72	51	0.56	40	0.04	3.1
	MC2	0.003	0.3	0.06	6.0	0.45	45	0.24	24	0.25	25
	MC3	0.006	0.7	0.09	10.5	0.45	52	0.30	35	0.02	2.2
NH$_4$-N	MC1	0.007	0.6	0.09	7.5	0.59	49	0.34	28	0.17	15
	MC2	0.005	0.3	0.10	6.1	0.61	36	0.44	26	0.53	31
	MC3	0.008	0.6	0.09	6.8	0.54	41	0.45	35	0.22	17
DON	MC1	0.01	1.0	0.15	12	0.53	44	0.29	24	0.23	19
	MC2	0.02	1.1	0.18	11	0.60	37	0.20	13	0.64	39
	MC3	0.01	0.6	0.16	7.2	0.90	41	0.82	38	0.28	13
P	MC1	0.009	6.0	0.05	32	0.05	32	0.05	30	0	0
	MC2	0.04	14	0.13	45	0.12	41	0	0	0	0
	MC3	0.04	17	0.14	61	0.05	22	0	0	0	0
S	MC1	0.03	1.5	0.28	13	1.1	51	0.64	30	0.10	4.5
	MC2	0.10	5.6	0.31	18	1.0	57	0.31	18	0.01	0.8
	MC3	0.04	3.6	0.24	21	0.65	57	0.20	18	0.007	0.5

Novel model approach. To assess the importance of stormflow for element export in our study catchments, we calculated the contribution of each flow class to element export by multiplying mean concentrations of chemical constituents in the flow classes with the 5-year mean annual discharge for the respective flow class. The mean five-year discharge of the five flow classes was derived from hourly flow rates modelled with TOPMODEL (Table D-2). The results were termed the "modeled export" (see Tables D-5 and D-6 for cumulative export rates for the 5-year period and Tables D-7 and D-8 for export per flow class). In our model, we assume that there is a "typical" mean concentration of each chemical constituent in each of the five flow classes as a result of flow conditions and soil properties in the flow region which is "active" during different flow classes. The small standard errors of the mean concentration of all chemical constituents in all flow classes (6-64% of the mean) and the lack of correlation between discharge rate and concentration within the five flow classes support this view.

Our new approach yielded total C, N, P, and S export rates which we considered sufficiently similar to those obtained from the FWM-based approach. The differences between the traditional and the modelled export rates were usually <20% for C, N, P, and S in all catchments except for P in MC1-3 and TOC in MC3. Phosphorus concentrations were consistently close to the detection limit in stream water thus limiting the accuracy of both calculation approaches. The larger difference in TOC export from MC3 between the two approaches might be related to the fact that the catchment sometimes received additional water from outside the usual catchment area during strong rainstorms as reflected by an extreme rise of the water levels on these occasions. The modelled temporal course does not represent these extreme high-flow events [*Fleischbein et al.*, 2006].

The agreement of the export rates calculated with our flow class-based model with the conventionally calculated flow weighted mean-based export rates was strong for K, Ca, Mg, and Na indicating that FWM-based export calculations are robust tools to assess base metal export (Table D-6). However, for Al and Mn differences between our and the conventional approach can be as much as 50% (Table D-6). This result indicates the problems of the FWM-approach in delivering good estimates of export rates of chemical constituents that mobilize strongly during extreme discharge events that are typically undersampled in the flow-weighted mean approach.

Table D-8: Mean element export of K, Ca, Mg, Na, Al, and Mn by discharge in different flow classes (April 1998- April 2003). Percentage values indicate the fraction of total export for the corresponding flow class.

	Catchment	Superdry kg ha^{-1} year^{-1}	%	Baseflow kg ha^{-1} year^{-1}	%	Intermediate kg ha^{-1} year^{-1}	%	Storm kg ha^{-1} year^{-1}	%	Lateral kg ha^{-1} year^{-1}	%
K	MC1	0.02	0.4	0.42	8.0	2.5	49	1.7	33	0.52	10
	MC2	0.04	0.9	0.45	11	2.1	52	0.77	19	0.71	17
	MC3	0.04	0.8	0.74	13	3.5	64	0.88	16	0.34	6.2
Ca	MC1	0.06	0.7	0.92	11	4.3	50	2.6	30	0.72	8.4
	MC2	0.09	1.2	0.78	10	4.0	52	1.5	19	1.3	17
	MC3	0.06	1.1	0.51	9.4	3.2	60	1.2	22	0.42	7.8
Mg	MC1	0.02	0.6	0.39	10	2.0	53	1.1	28	0.32	8.4
	MC2	0.05	1.1	0.53	11	2.7	54	0.89	18	0.84	17
	MC3	0.05	1.2	0.36	9.8	2.2	60	0.84	23	0.21	5.8
Na	MC1	0.32	1.0	4.1	13	17	54	8.0	26	1.8	5.7
	MC2	0.77	2.4	4.3	14	17	55	6.0	19	3.3	11
	MC3	0.37	1.9	2.8	14	13	64	3.4	17	0.67	3.4
Al	MC1	0.001	0.6	0.009	5.7	0.028	19	0.017	11	0.099	64
	MC2	0.002	0.6	0.016	6.2	0.048	19	0.015	6.1	0.17	68
	MC3	0.002	0.3	0.015	1.7	0.21	22	0.44	48	0.26	28
Mn	MC1	0.0001	0.6	0.001	17	0.002	36	0.0003	5.1	0.003	42
	MC2	0	0	0.001	11	0.004	33	0.001	9.0	0.005	47
	MC3	0.0001	0.2	0.001	3.2	0.001	33	0.013	45	0.006	19

4.4 Hypothesis 4: Storm events have a significant influence on catchment nutrient export

C, N, P, and S. For P and S, peak flow-related export was of little importance (<20% of the total export occurred during peak discharge conditions, i.e. sum of flow classes "stormflow" and "lateral flow", Table D-7) except for MC1 (30%). In contrast to P and S, peak discharge accounted for approximately 50% of the export of TOC and the N species (Table D-7). Similar results were reported by Buffam *et al.* [2001], who found that more than 50% of the DOC, DON, and NO_3-N export were associated with stormflow in an Appalachian stream. For TOC, NH_4-N, and DON, the rarely occurring flow class "lateral flow" contributed as much to the export as did the more frequently occurring flow class "stormflow" (Table D-5). Thus, these results support our fourth hypothesis concerning C and N export but not concerning P and S export.

Metals. The extreme flow classes "storm" and "lateral" accounted for an estimated 50-75% of the total export of Al and Mn (Table D-8). These two flow classes also included around one third of the total export of K, Ca, Mg, and Na (Table D-8). Our values are in good agreement with those obtained by Grimaldi *et al.* [2004] and Stoorvogel *et al.* [1997]. In another study on export rates of catchments with non-volcanic geology in Malaysia, Yusop *et al.* [2006] reported even greater contributions of stormflow to the export rates of K (55-72%), Ca (43-62%), and Mg (44-56%).

Greater metal export during stormflow than during baseflow was also reported for catchments on deeply weathered metamorphic sandstones for Ca, Mg, and K in Western Amazonia and Malaysia [*Elsenbeer et al.*, 1996; *Yusop et al.*, 2006] and on amphibolites in Queensland for K [*Elsenbeer et al.*, 1994] and Al and Mn [*Lorieri and Elsenbeer*, 1997]. On volcanic bedrock, in contrast, metal export rose with decreasing discharge as reported for Ca and Mg in French Guiana [*Grimaldi et al.*, 2004] and for Ca, Mg, K, and Na in Puerto Rico [*Schellekens et al.*, 2004].

The role of stormflow and lateral flow as a major pathway for C, N, Al and Mn export and as a considerable export pathway for base metals might be of growing importance because Hinton *et al.* [1998] predict changes in the frequency of storm events. Haylock *et al.* [2006] found a trend towards wetter and more extreme rainfall conditions for Ecuador by analyzing rainfall data from 1960 to 2000, and explained this by changes of the El Niño Southern Oscillation (ENSO).

5. Conclusions

In three microcatchments under montane forest in Ecuador, concentrations of chemical constituents in stream water were different among different discharge classes but there was no correlation between discharge rates and concentrations. Carbon, NO_3-N, NH_4-N, DON, K, Ca, Mg, Al and Mn showed positive, P, S and Na negative responses to increase in discharge. Stormflow and lateral flow flushed all chemical constituents except for P and Na from the element-rich topsoil by rapid near-surface flow. Total organic C, DON, Al, and Mn concentrations in stream water were particularly large when rainstorms coincided with waterlogging of soils thereby promoting near-surface lateral flow in the organic layer. Phosphorus and Na, in contrast, were mainly released into the stream water by weathering of the subsoil which was also the most important S source to the stream.

Metal concentrations in stream water were only explained to a small extent by vertical metal distribution. Controlling processes of metal concentrations were pH and organo-complexation, which varied with water flow depth in soil. Decreasing pH in soil water caused increasing solubilization of metals. The degree of organo-complexation of metals increased with increasing TOC concentrations in stream water and was also related to flow depth through the soil, in which organic matter concentrations decreased with increasing depth.

Because of the lack of a correlation between discharge rates and concentrations of chemical constituents in stream water and as a consequence of our sampling scheme in regular intervals, we developed a novel tool to classify flow classes and to model the contributions of different flow regimes to total element export. This new approach was successfully validated by comparison with the traditional FWM-based approach and proved useful to quantify the contributions of different flow regimes to total element export.

Storm events had a significant influence on C, N, and metal export of tropical montane forest. About 50% of the total TOC, NH_4-N and DON export and about 40% of the total NO_3-N export were caused by storm event-related discharge (i.e. sum of our flow classes lateral flow and storm flow). Storm events also accounted for roughly 30% of total base metal (K, Ca, Mg, Na) export and for up to 75% of total Al and Mn export. In spite of the rarity of storm events with antecedent water saturation in subsoil, strong near-surface lateral flow is an important export pathway for TOC, N and metal export in steep, forested catchments in south Ecuador. Storm event-related export of S was less (around a fifth of total export), yet not negligible. An accurate calculation of element export therefore requires the careful consideration of storm events.

With the exception of Na and Al, the exported elements monitored are nutrients that are locally poor in supply for Amazonian montane forest ecosystems. Therefore, the proposed global-warming induced frequency shift in extreme rain events may cause nutrient depletion of Amazonian montane forest ecosystems.

6. Acknowledgments

We are particularly grateful to Wolfgang Zech who initiated this long-term research project. We thank Paul Emck for providing meteorological data, Syafrimen Yasin for the soil data and Katrin Fleischbein and Rainer Goller for providing surface flow and chemical concentration data and many student helpers for their support. All P and S species analyses were performed at the Central Analytic Department of the Bayreuther Institut für Terrestrische Ökosystemforschung (BITÖK).

We are grateful to Nature and Culture International (NCI) in Loja, Ecuador for providing the study area and the research station and to the Instituto Ecuatoriano Forestal de Areas Naturales y Vida Silvestre (INEFAN) for the permission to conduct this study. This investigation was funded by the Deutsche Forschungsgemeinschaft (FOR402, Wi1601/5-2, -3).

7. References

Beven, K. J., R. Lamb, P. F. Quinn, and R. B. Romanowicz (1995), TOPMODEL, in *Computer Models of Watershed Hydrology*, edited by V. P. Singh, pp. 627-668, Water Resource Publications, Colorado.

Brown, V. A., J. J. McDonnell, D. A. Burns, and C. Kendall (1999), The role of event water, a rapid shallow flow component, and catchment size in summer stormflow, *Journal of Hydrology*, *217*(3-4), 171-190.

Bruijnzeel, L. A., and L. S. Hamilton (2000), Decision time for cloud forests, 1-41 pp, IHP-Unesco and WWF International, Paris, Amsterdam.

Buffam, I., J. N. Galloway, L. K. Blum, and K. J. McGlathery (2001), A stormflow/baseflow comparison of dissolved organic matter concentrations and bioavailability in an Appalachian stream, *Biogeochemistry*, *53*(3), 269-306.

Campbell, J. L., J. W. Hornbeck, W. H. McDowell, D. C. Buso, J. B. Shanley, and G. E. Likens (2000), Dissolved organic nitrogen budgets for upland, forested ecosystems in New England, *Biogeochemistry*, *49*(2), 123-142.

Campo, J., J. M. Maass, V. J. Jaramillo, and A. M Yrizar (2000), Calcium, potassium, and magnesium cycling in a Mexican tropical dry forest ecosystem, *Biogeochemistry*, *49*(1), 21-36.

Cuevas, E., and E. Medina (1988), Nutrient dynamics within amazonian forests II. Fine root growth, nutrient availability and litter decomposition, *Oecologia*, *76*, 222-235.

Elsenbeer, H., A. West, and M. Bonell (1994), Hydrologic Pathways and Stormflow Hydrochemistry at South Creek, Northeast Queensland, *Journal of Hydrology*, *162*(1-2), 1-21.

Elsenbeer, H., A. Lack, and K. Cassel (1996), The stormflow chemistry at La Cuenca, Western Amazonia, *Interciencia*, *21*(3), 133-139.

Elsenbeer, H. (2001), Hydrologic flowpaths in tropical rainforest soilscapes - a review (vol 15, pg 1755, 2001), *Hydrological Processes*, *15*(14), 2863-2863.

Evans, C., and T. D. Davies (1998), Causes of concentration/discharge hysteresis and its potential as a tool for analysis of episode hydrochemistry, *Water Resources Research*, *34*(1), 129-137.

Fleischbein, K., W. Wilcke, R. Goller, J. Boy, C. Valarezo, W. Zech, and K. Knoblich (2005), Rainfall interception in a lower montane forest in Ecuador: effects of canopy properties, *Hydrological Processes*, *19*(7), 1355-1371.

Fleischbein, K., W. Wilcke, C. Valarezo, W. Zech, and K. Knoblich (2006), Water budgets of three small catchments under montane forest in Ecuador: experimental and modelling approach, *Hydrological Processes*, *20*(12), 2491-2507.

Genereux, D. (2004), Comparison of naturally-occurring chloride and oxygen-18 as tracers of interbasin groundwater transfer in lowland rainforest, Costa Rica, *Journal of Hydrology*, *295*(1-4), 17-27.

Godsey, S., H. Elsenbeer, and R. Stallard (2004), Overland flow generation in two lithologically distinct rainforest catchments, *Journal of Hydrology*, *295*(1-4), 276-290.

Goller, R., W. Wilcke, M. J. Leng, H. J. Tobschall, K. Wagner, C. Valarezo, and W. Zech (2005), Tracing water paths through small catchments under a tropical montane rain forest in south Ecuador by an oxygen isotope approach, *Journal of Hydrology*, *308*(1-4), 67-80.

Goller, R., W. Wilcke, K. Fleischbein, C. Valarezo, and W. Zech (2006), Dissolved nitrogen, phosphorus, and sulfur forms in the ecosystem fluxes of a montane forest in ecuador, *Biogeochemistry*, *77*(1), 57-89.

Grimaldi, C., M. Grimaldi, A. Millet, T. Bariac, and J. Boulegue (2004), Behaviour of chemical solutes during a storm in a rainforested headwater catchment, *Hydrological Processes*, *18*(1), 93-106.

Gustafsson, J. P., D. Berggren, M. Simonsson, M. Zysset, and J. Mulder (2001), Aluminium solubility mechanisms in moderately acid Bs horizons of podzolized soils, *European Journal of Soil Science*, *52*(4), 655-665.

Hagedorn, F., P. Schleppi, P. Waldner, and H. Fluhler (2000), Export of dissolved organic carbon and nitrogen from Gleysol dominated catchments - the significance of water flow paths, *Biogeochemistry*, *50*(2), 137-161.

Haylock, M. R., et al. (2006), Trends in total and extreme South American rainfall in 1960-2000 and links with sea surface temperature, *Journal of Climate*, *19*(8), 1490-1512.

Hedin, L. O., P. M. Vitousek, and P. A. Matson (2003), Nutrient losses over four million years of tropical forest development, *Ecology*, *84*(9), 2231-2255.

Hinton, M. J., S. L. Schiff, and M. C. English (1998), Sources and flowpaths of dissolved organic carbon during storms in two forested watersheds of the Precambrian Shield, *Biogeochemistry*, *41*(2), 175-197.

Homeier, J. (2004), Baumdiversität, Waldstruktur und Wachstumsdynamik zweier tropischer Bergregenwälder in Ecuador und Costa Rica, Ph. D thesis, University of Bielefeld, Bielefeld, Germany.

Hook, A. M., and J. A. Yeakley (2005), Stormflow dynamics of dissolved organic carbon and total dissolved nitrogen in a small urban watershed, *Biogeochemistry*, *75*(3), 409-431.

Kaiser, K., G. Guggenberger, and W. Zech (2001), Organically bound nutrients in dissolved organic matter fractions in seepage and pore water of weakly developed forest soils, *Acta Hydrochimica Et Hydrobiologica*, *28*(7), 411-419.

Lewis, W. M., J. M. Melack, W. H. McDowell, M. McClain, and J. E. Richey (1999), Nitrogen yields from undisturbed watersheds in the Americas, *Biogeochemistry*, *46*(1-3), 149-162.

Likens, G. E., and J. S. Eaton (1970), A polyurethane stemflow collector for trees and shrubs, *Ecology*, *51*, 938-939.

Likens, G. E., C. T. Driscoll, D. C. Buso, M. J. Mitchell, G. M. Lovett, S. W. Bailey, T. G. Siccama, W. A. Reiners, and C. Alewell (2002), The biogeochemistry of sulfur at Hubbard Brook, *Biogeochemistry*, *60*(3), 235-316.

Liu, W. Y., J. E. D. Fox, and Z. F. Xu (2003), Nutrient budget of a montane evergreen broad-leaved forest at Ailao Mountain National Nature Reserve, Yunnan, southwest China, *Hydrological Processes*, *17*(6), 1119-1134.

Lloyd, C. R., and A. Marques (1988), Spatial variability of throughfall and stemflow measurements in Amazonian rain forest, *Agricultural and Forest Meteorology*, *47*, 63-73.

Lorieri, D., and H. Elsenbeer (1997), Aluminium, iron and manganese in near-surface waters of a tropical rainforest ecosystem, *Science of the Total Environment*, *205*(1), 13-23.

Martinelli, L. A., M. C. Piccolo, A. R. Townsend, P. M. Vitousek, E. Cuevas, W. McDowell, G. P. Robertson, O. C. Santos, and K. Treseder (1999), Nitrogen stable isotopic composition of leaves and soil: Tropical versus temperate forests, *Biogeochemistry*, *46*(1-3), 45-65.

Matzner, E. (2004), *Biogeochemistry of Forested Catchments in a changing Environment. A German case study*, Springer, Berlin.

McDowell, W. H., and C. E. Asbury (1994), Export of Carbon, Nitrogen, and Major Ions from 3 Tropical Montane Watersheds, *Limnology and Oceanography*, *39*(1), 111-125.

Michalzik, B., K. Kalbitz, J. H. Park, S. Solinger, and E. Matzner (2001), Fluxes and concentrations of dissolved organic carbon and nitrogen - a synthesis for temperate forests, *Biogeochemistry*, *52*(2), 173-205.

Mitchell, M. J., B. Mayer, S. W. Bailey, J. W. Hornbeck, C. Alewell, C. T. Driscoll, and G. E. Likens (2001), Use of stable isotope ratios for evaluating sulfur sources and losses at the Hubbard Brook Experimental Forest, *Water Air and Soil Pollution*, *130*(1-4), 75-86.

Möller, A., K. Kaiser, and G. Guggenberger (2005), Dissolved organic carbon and nitrogen in precipitation, throughfall, soil solution, and stream water of the tropical highlands in northern Thailand, *Journal of Plant Nutrition and Soil Science-Zeitschrift Fur Pflanzenernahrung Und Bodenkunde*, *168*(5), 649-659.

Oyarzun, C. E., R. Godoy, A. De Schrijver, J. Staelens, and N. Lust (2004), Water chemistry and nutrient budgets in an undisturbed evergreen rainforest of southern chile, *Biogeochemistry*, *71*(1), 107-123.

Parker, G. G. (1983), Throughfall and stemflow in the forest nutrition cycle, *Advances in Ecological Research*, *13*, 57-133.

Perakis, S. S., and L. O. Hedin (2002), Nitrogen loss from unpolluted South American forests mainly via dissolved organic compounds, *Nature*, *415*(6870), 416-419.

Richter, M. (2003), Using epiphytes and soil temperatures for eco-climatic interpretations in Southern Ecuador, *Erdkunde*, *57*, 161-181.

Rieuwerts, J. S. (2007), The mobility and bioavailability of trace metals in tropical soils: a review, *Chemical Speciation and Bioavailability*, *19*(2), 75-85.

Rollenbeck, R., J. Bendix, P. Fabian, J. Boy, H. Dalitz, P. Emck, M. Oesker, and W. Wilcke (2007), Comparison of different techniques for the measurement of precipitation in tropical montane rain forest regions, *Journal of Atmospheric and Oceanic Technology*, *24*(2), 156-168.

Saunders, T. J., M. E. McClain, and C. A. Llerena (2006), The biogeochemistry of dissolved nitrogen, phosphorus, and organic carbon along terrestrial-aquatic flowpaths of a montane headwater catchment in the Peruvian Amazon, *Hydrological Processes, 20*(12), 2549-2562.

Schellekens, J., F. N. Scatena, L. A. Bruijnzeel, A. I. J. M. van Dijk, M. M. A. Groen, and R. J. P. van Hogezand (2004), Stormflow generation in a small rainforest catchment in the luquillo experimental forest, Puerto Rico, *Hydrological Processes, 18*(3), 505-530.

Schrumpf, M., G. Guggenberger, C. Schubert, C. Valarezo, and W. Zech (2001), Tropical montane forest soils: development and nutrient status along an altitudinal gradient in the south Ecuadorian Andes, *Die Erde, 132*, 43-59.

Stoorvogel, J. J., B. H. Janssen, and N. VanBreemen (1997), The nutrient budgets of a watershed and its forest ecosystem in the Tai National Park in Cote d'Ivoire, *Biogeochemistry, 37*(2), 159-172.

Tromp-van Meerveld, H. J., and J. J. McDonnell (2006), Threshold relations in subsurface stormflow: 2. The fill and spill hypothesis, *Water Resources Research, 42*(2), W02411.

Wardle, D. A., L. R. Walker, and R. D. Bardgett (2004), Ecosystem properties and forest decline in contrasting long-term chronosequences, *Science, 305*(5683), 509-513.

Western, A. W., S. L. Zhou, R. B. Grayson, T. A. McMahon, G. Bloschl, and D. J. Wilson (2004), Spatial correlation of soil moisture in small catchments and its relationship to dominant spatial hydrological processes, *Journal of Hydrology, 286*(1-4), 113-134.

Wilcke, W., S. Yasin, C. Valarezo, and W. Zech (2001), Change in water quality during the passage through a tropical montane rain forest in Ecuador, *Biogeochemistry, 55*(1), 45-72.

Wilcke, W., S. Yasin, U. Abramowski, C. Valarezo, and W. Zech (2002), Nutrient storage and turnover in organic layers under tropical montane rain forest in Ecuador, *European Journal of Soil Science, 53*(1), 15-27.

Yusop, Z., I. Douglas, and A. R. Nik (2006), Export of dissolved and undissolved nutrients from forested catchments in Peninsular Malaysia, *Forest Ecology and Management, 224*(1-2), 26-44.

Die VDM Verlagsservicegesellschaft sucht für wissenschaftliche Verlage abgeschlossene und herausragende

Dissertationen, Habilitationen, Diplomarbeiten, Master Theses, Magisterarbeiten usw.

für die kostenlose Publikation als Fachbuch.

Sie verfügen über eine Arbeit, die hohen inhaltlichen und formalen Ansprüchen genügt, und haben Interesse an einer honorarvergüteten Publikation?

Dann senden Sie bitte erste Informationen über sich und Ihre Arbeit per Email an *info@vdm-vsg.de*.

Sie erhalten kurzfristig unser Feedback!

VDM Verlagsservicegesellschaft mbH
Dudweiler Landstr. 99   Telefon +49 681 3720 174
D - 66123 Saarbrücken   Fax   +49 681 3720 1749
www.vdm-vsg.de

Die VDM Verlagsservicegesellschaft mbH vertritt

Printed by Books on Demand GmbH, Norderstedt / Germany